fP

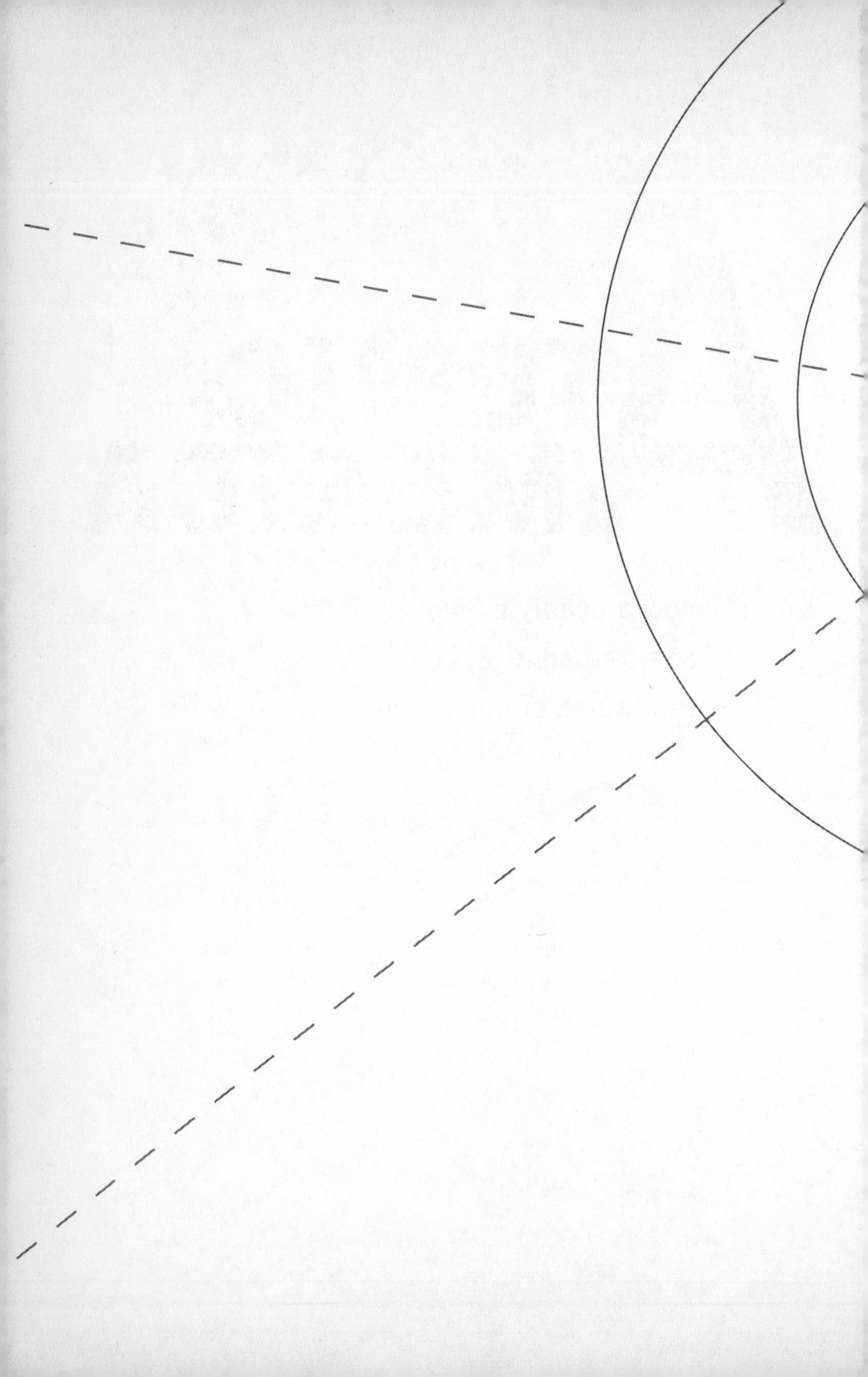

ATOMIC AMERICA

how a deadly explosion and a feared admiral changed the course of nuclear history

TODD TUCKER

FREE PRESS
new york london toronto sydney

*f*P

FREE PRESS
A Division of Simon & Schuster, Inc.
1230 Avenue of the Americas
New York, NY 10020

First Free Press hardcover edition March 2009

FREE PRESS and colophon are trademarks of Simon & Schuster, Inc.

For information about special discounts for bulk purchases, please contact Simon & Schuster Special Sales at 1-800-456-6798 or business@simonandschuster.com

Designed by Suet Y. Chong

Manufactured in the United States of America

10 9 8 7 6 5 4 3 2 1

Library of Congress Cataloging-in-Publication Data

Tucker, Todd
Atomic America : how a deadly explosion and a feared admiral changed the course of nuclear history / by Todd Tucker.—1st Free Press hardcover ed.
p. cm.
Includes bibliographical references and index.
1. Nuclear reactor accidents—Idaho—Idaho Falls—History—20th century.
2. Nuclear Industry—Military aspects—United States—History—20th century.
3. Idaho Falls (Idaho)—History—20th century. I. Title.
TK1345.I2T83 2009
363.17'990979653—dc22 2008013842

ISBN-13: 978-1-4165-4433-3
ISBN-10: 1-4165-4433-X

Dedicated to
John Byrnes, Richard Legg, and Richard McKinley

CONTENTS

Supercritical: The status of a reactor that has proceeded past the just-critical point, generating one new fission per previous fission. The added fissions increase exponentially, power going up in the same proportion. If not contained with control rods, the reactor will run away.

—*Nuclear Flight: The United States Air Force Programs for Atomic Jets, Missiles, and Rockets,*
Lieutenant Colonel Kenneth F. Gantz, USAF, editor, 1960

PROLOGUE
JANUARY 3, 1961

A moment before his death, John Byrnes knelt atop the Army's SL-1 reactor, poised to pull the central control rod straight up. His supervisor, Richard Legg, was nearby. The third crewman, Richard McKinley, was pacing around the vessel head, between the movable shield blocks and the motor control panel. As the newest member of the cadre, just three weeks into his hitch in Idaho, McKinley was probably running tools and trying to learn what he could between errands. He must also have observed the simmering tension between his two crewmates.

Legg and Byrnes had arrived in Idaho together, in October 1959, and had clashed since those first days. They had even come to drunken blows at a sleazy bachelor party the year before. But Legg had since surpassed Byrnes professionally and qualified as both chief operator and shift supervisor—this was Byrnes's first shift as Legg's subordinate. Byrnes's steady record of disciplinary problems all but guaranteed that his professional progress in the Army was over. Byrnes hated Legg.

The desolation surrounding them would have reinforced a dark mood, a landscape where even the place-names evoked solitude and despair. The Lost River Desert, the Snake River Plain, and the Craters of the Moon were all places the drafty government buses drove them through on their daily hundred-mile round-trip to the reactor. Much of the ground was covered in ancient black lava so hard and so thick that site engineers had trouble blasting through it even with shaped charges of dynamite as they busily erected experimental reactors up and down the plain. And January 3 was bitterly cold—the overnight low in Idaho Falls was six degrees below zero. Over the decades as the story was retold, many would recall it being even colder.

The reactor that Byrnes, McKinley, and Legg worked on was unglamorous and unloved even inside the fences of the National Reactor Testing Station in Idaho. The Navy reactors, in contrast, run by the brilliant and tyrannical Admiral Hyman Rickover, were the pride of the base. Just three years earlier, Rickover had stunned the world when the nuclear-powered USS Nautilus *had steamed under the North Pole. It was the stuff of Jules Verne, a development that promised to change the nature of warfare: a submarine that could stay submerged forever. The prototype for that reactor, S1-W, operated in a giant tank of water to simulate the submarine environment, just ten miles northwest of SL-1 but worlds away in terms of prestige and excitement. On the northern end of the sprawling Idaho reservation, jealous Air Force generals played catch-up, pouring hundreds of millions of dollars into a nuclear-powered jet airplane, a giant bomber that would stay aloft for years, if they could ever get the behemoth off the ground.*

The Army's goal for nuclear power was vastly more prosaic: small, semiportable power stations for remote bases. Of the more than twenty reactors in Idaho, SL-1 was the smallest, designed merely to generate 200 kilowatts of electricity.

Professional disappointment was just one of many reasons the volatile Jack Byrnes might have been distracted that cold night. He was probably exhausted, having slept on friends' couches the previous two nights, as the latest fight between him and wife, Arlene Byrnes, ran its course. The fight had come at the end of their too-short holiday break, and Byrnes had returned to SL-1 to find a long list of maintenance he was supposed to complete under Legg's supervision, a list that ended with the start-up of the troublesome little reactor. Five hours into the watch, they had barely completed anything.

At 7:00 PM, Arlene had called SL-1 and told Jack that she wanted a divorce. After a year of fighting and loneliness in the Lost River Desert, Arlene Byrnes had finally had enough. Their last conversation ended with a discussion of how to split his paltry Army paycheck.

So at 9:00, it may have been difficult for Byrnes to focus on the task at hand. The procedure for reassembling the control rod drive mechanism called for lifting the rod "not more than four inches." Byrnes was no nuclear engineer, but he was a well-trained Army specialist—he knew that the central rod in SL-1, by virtue of its position in the core, was enormously powerful, capable of starting up the reactor all by itself. If having his hands on that rod wasn't nerve-wracking enough, Byrnes might also have been uneasy to have Legg hovering so closely behind him. Self-conscious about his height at five foot six inches, Legg was constantly physically asserting himself, challenging any and all to wrestling matches and goosing his comrades at inappropriate times. Hunched over the control rod, straining with effort, Byrnes would have made a tempting target for one of Legg's pranks. And Byrnes's task would not have been easy, even without Legg looming behind him. The rod was heavy: eighty-four pounds. What's more, the boron strips inside the core were crumbling, occasionally jamming the control rods in their channels and making them almost impossible to move, a problem that had gotten worse in recent months. Sometimes even the drive motors couldn't move the rods, and old-fashioned Army ingenuity would be applied to the problem, usually in the form of a pipe wrench.

At 9:00 PM, three hours remained in the shift, three hours that must have stretched out like an eternity before Jack Byrnes. There were many things that might have been running through his exhausted mind—perhaps even the terse warnings of the procedure he was about to perform. Despite four decades of speculation, however, no one will ever know exactly what he was thinking at the moment he tightened his hands around the rod, and pulled.

• • •

At 9:01 PM, January 3, 1961, a nuclear reactor exploded in Idaho, killing three men who now lie buried in lead-lined caskets. It remains the only fatal reactor accident in American history.

The details released to the newspapers immediately after the explosion were deliberately vague, not so much because of Cold War secrecy, but more in an effort to spare the three widows the gruesome details of their husbands' deaths. The interim accident report published in May 1961 by the Atomic Energy Commission was less coy, as it straightforwardly described the position of the three radioactive bodies immediately after the explosion:

> *The #2 crew member was struck on his back and legs with water and/or steam causing him to be thrown against a shield block and landing in the vicinity of the instrument wells. The #1 crew member was also struck with water and/or steam and was thrown back against another shield block striking his head first. Simultaneously, the No. 7 shield plug assembly impaled the #3 crew member and pinned him to the bottom of the fan floor a distance of approximately 13 feet above the reactor head.*

The #3 crew member, Richard Legg, had been standing over the rod 7 plug assembly when the explosion occurred. The plug assembly was a metal shaft placed over the control rod, but it was not the control rod itself that impaled Legg, as was often stated later. The shield plug was ejected from the core at eighty-five feet per second, entered Legg's body through his groin, exited near his shoulder, and propelled him straight up to the ceiling where he dangled for six days. The impaled body was so radioactive that it took engineers that long to design a safe way to remove it. When they did finally bring Legg down, they were shocked to see that despite the time that had passed, the body was perfectly preserved. It was so radioactive that the sterilized flesh had not decayed.

Nuclear power was the younger sibling of the atomic bomb, and both were children of the Manhattan Project. The first nuclear reactors had been a means to an end, the production of plutonium for weapons. After the war, among the scientists and engineers who designed the bomb there was an almost spiritual desire to create something productive from their monumental work, something that would balance the tremendous destructive power they had unleashed over Hiroshima and Nagasaki. If the atomic bomb was the ultimate weapon, a risk to civilization itself, then atomic energy must be an energy source of unlimited beneficence, the power to uplift all of mankind. Billions of dollars would be spent to prove it true.

But SL-1 was a military reactor, as nuclear power in its infancy was almost exclusively a military enterprise. In those early days, only federal dollars and the urgency of military requirements could support the vast investment necessary to make nuclear power a reality. In addition, the line between nuclear power and nuclear weaponry was blurry, just as it is now, making the military reluctant to relinquish its hold on the nation's nuclear reactors, no matter how often the spirit of "Atoms for Peace" was invoked. Each military service made the case that it urgently needed nuclear power. The Army wanted portable, tireless power plants for Arctic radar bases, the first line of defense against a Soviet air attack. The Air Force wanted a supersonic bomber with unlimited endurance, the ultimate weapon in a world where airpower was ascendant. And the Navy wanted to fulfill the dream of a "true submarine," a ship that would live beneath the waves. Each service was convinced that without perfecting a mission for the Atomic Age, it would become obsolete. Interservice rivalry is a grand American tradition, but in those tense early days of the Cold War, the stakes had never been higher.

The explosion at SL-1 led to the end of the Army program, happened within weeks of the end of the Air Force's atomic plane, and opened the door for the Navy's long-standing, jealously guarded monopoly on military nuclear power. The civilian industry has for more than a generation been staffed largely by Navy veterans, and the Navy philosophy has, in large part, become the industry's philosophy. On March 28, 1979, Three Mile Island became a virtual brand name for nuclear disaster, resulting in showy but shallow reforms. SL-1 affected the DNA of the industry in utero, transforming the very philosophy of nuclear engineering. Questions as fundamental as the number of control rods necessary to run a plant safely were settled at SL-1. The dream of miniaturized, portable nuclear plants died with McKinley, Legg, and Byrnes.

Today, people on both sides of the nuclear power debate anticipate an atomic renaissance in America. Chronic instability in the Middle East and fears of global warming have brought together an unlikely coalition of conservative isolationists and fretful environmentalists, all of whom argue that emission-free nuclear power, already quietly responsible for nearly 20 percent of the American energy supply, is ready to take on more of the nation's energy burden. Over one hundred nuclear plants generate electricity in the United States. Thirty-two new reactors are planned. All are descendants of SL-1. With so many plants in operation, and so many more on the way, it is vitally important to understand the real reasons, technical and otherwise, a nuclear reactor exploded in 1961.

Understanding the dominant theory of the explosion does not require a familiarity with nuclear physics. Within days of the incident, rumors of infidelity and a love triangle sprang to life, stories of an aggrieved husband who used a control rod as a peculiarly modern murder weapon. There was just enough truth in the story, and it explained the tragedy so neatly, that it rapidly obscured the underlying issues at SL-1. One of the young crewmen was, in fact, unstable and had a collapsing marriage. But why was he allowed to perform dangerous maintenance with so little supervision? The rapid, manual raising of a single control rod did cause the devastation at SL-1. But why would a reactor be designed so perilously close to criticality? Why would procedures actually dictate the manual lifting of that rod? The story of the love triangle and murder-suicide has been handed down faithfully by people in Idaho, the military, and the nuclear industry. Like all good folklore, it embodies the bedrock principles of those who keep it alive, even, or perhaps especially, where it diverges from the facts. And while the story has proven to be extremely durable, it is an inadequate description of what really caused America's only fatal reactor accident, and what lessons should be learned.

The complete story of SL-1 is neither a murder mystery nor a love story. It is more than an engineering case study as well, an incident that cannot be explained completely with flux diagrams and reactivity calculations. The story of SL-1 is a war story, a tale of a bloody, costly struggle between the three branches of the United States military. It happened at a time when nuclear annihilation was a frighteningly real possibility: January 3, 1961, was eight months after Francis Gary Powers was shot down over the Soviet Union, and three months before the Bay of Pigs invasion. The military was preparing to fight a nuclear world war that many viewed as inevitable, and nuclear power was seen by each branch as a way, quite literally, to increase its power. Massive budgets were at stake, but it was more than that. The generals and admirals all believed passionately that survival was at stake both for their service, and for the nation. In this struggle for nuclear supremacy, the Army skirmished on the fringes, but the Navy and the Air Force were in a fight to the death, as the flyers argued convincingly that airpower and only airpower could save America from the communist menace. And while many of this war's battles took place in Idaho, it began, as so many American wars have, with an attack on a ship.

chapter 1

THE USS *UNITED STATES*

While the Manhattan Project was a jealously guarded Army protectorate, the Navy somehow managed to insert two of its officers into pivotal roles: both atomic bombs dropped on Japan were armed in flight by U.S. naval officers. Captain William "Deak" Parsons, USN, climbed into the hold of the *Enola Gay* and armed Little Boy en route to Hiroshima, and Commander Frederick Ashworth did the same for Fat Man on the way to Nagasaki. After their B-29s banked away from the atomic explosion, after they watched the sky boil and felt the impact of the pressure wave and its reflection, those two men may have become the first acolytes of what would become pillars of naval doctrine in the postwar years: the future of warfare was atomic. And the future of the Navy was the USS *United States*.

Her hull number would be CVA-58, the "CV" the Navy's standard designation for a fleet aircraft carrier, and the "A" a new designator for "atomic." The *United States* was "atomic" because she was large enough to carry nuclear bombs and the planes that could drop them. At seventy-nine thousand tons, she would be the largest warship ever built by any navy, 50 percent larger than the *Midway*, the largest carrier of World War II, and more than double the size of the *Essex* class carriers, the workhorse of the fleet and the carriers that won the war in the Pacific. Her size was dictated by the huge planes she was to carry, which were in turn engineered around the massive weight of the atomic weapons of the era. The "Fat Man" style bombs weighed ten thousand pounds apiece, and required hundred-thousand-pound airplanes to carry them. One carrier-launched bomber of the Navy in World War II, in comparison, was the SB2C Helldiver. It weighed a sprightly thirteen thousand pounds. In the nuclear age, everything needed to be bigger.

Another point of comparison: the Helldiver was thirty-six feet long and had a wingspan of forty-nine feet. The Navy theorized that its new atomic bomber, called ADR-42 on the drawing boards, would be eighty-seven feet long and have a wingspan of 110 feet. That massive wingspan could never work on a traditional carrier; during takeoff and landing the wings could never get by the "island," the structure that rose from a carrier's flight deck, containing the bridge, the radar masts, and the exhaust stacks that made operating an aircraft carrier possible. The presence of the island limited the wingspan of carrier-borne aircraft, and this limitation led to the most radical aspect of the new carrier's design: her deck would be completely flush.

Radar and communications would be provided by other ships. Exhaust gases from the engine room would pass not through smokestacks, but through gill-like openings on her sides. Since the *United States,* as huge as she was, would only be able to carry a limited number of the bombers and their atomic weapons, the ship's defense would also be the responsibility of other ships. It would require a whole battle group to support and defend her, including at least two other conventionally designed aircraft carriers to carry fighter aircraft. The entire concept—a huge carrier, new giant bombers, hundreds of atomic bombs, a dozen or more ships in support—represented an unprecedented investment, especially in peacetime. No matter. The new ship would allow the nation to project its monopoly on atomic power to any corner of the globe. And the USS *United States* wasn't just revolutionary; she was beautiful. With no protrusions marring the flight deck or limiting wingspans, she was the ultimate realization of carrier design, a huge, pure runway upon the seas.

The total cost of the project was staggering. The *United States* by herself would cost $189 million, and the total cost of her dedicated battle group might run as high as $800 million—a mind-boggling amount of money in 1949, when the postwar nation was eagerly slashing defense budgets. The Navy, religiously committed to the project, agreed to cancel the construction of thirteen other ships to offset some of the cost, and, along with constant, zealous lobbying, it managed to get the budget for the massive ship approved. Construction would take place in Shipway No. 11 at Newport News Shipbuilding in Virginia. During the war, the massive Shipway No. 11 had made six LSTs at a time, the giant ships designed to ferry tanks and troops from shore to shore during amphibious landings.

The USS *United States* would not only fill the shipway by herself, but would also rise above and swell beyond the dock as she took shape during her four-year construction. She was so big it took shipyard workers a full month just to place the wooden blocks in the bottom of the dry dock that would hold her keel plates in position while they were welded together. Finally, at 8:45 AM on a cold, drizzly Monday, April 18, 1949, the largest crane at Newport News carefully lowered into the bottom of Shipway No. 11 a flat plate of steel, twenty-four by thirty feet square and one inch thick, the first fifteen tons of a ship that would weigh seventy-nine thousand tons. Someone must have pronounced the keel "well and truly laid," in accordance with naval tradition, and who present could doubt it? They were witnessing history, the ascendance of the USS *United States* as the world's greatest ship, and the United States as the world's greatest maritime power.

Five days later, construction of the *United States* was canceled.

• • •

It was not, as later claimed in Navy legend, a complete sneak attack. The Newport News *Times-Herald*, as the main paper in a company town, was ever sensitive to the vagaries of federal contracts. The headline even on the afternoon of the keel laying warned readers that "Controversy Still Rages on Capitol Hill Over Craft." In fact, the Navy seemed to be hurriedly laying the keel in the hope that doing so would make the ship a fait accompli, no longer subject to cancellation. The keel laying for the *United States* took place without ceremony or dignitaries, unusual for any large warship, conspicuous for the largest warship ever made. Neither the president of the shipyard nor the mayor of Newport News, R. Cowles Taylor, attended the event, although the mayor did find the time to react indignantly to the news that Elmer Davis, a nationally syndicated radio announcer, had erroneously broadcasted to the nation that the keel laying took place in Norfolk. The local paper summed up the position of the town and the Navy when it wrote hopefully, "the facts in the matter are that the keel has been laid, steel fabrication is well under way and opponents of the plan will have a harder time now trying to get the contract thrown out."

The "opponents of the plan" were the Navy's most dangerous natural enemy: the U.S. Air Force. Although an independent entity only since 1947, the Air Force had done a masterful job of convincing powerful people in the Capitol, the Pentagon, and the public that nuclear weapons had made the Navy, and to a lesser extent the Army, obsolete. Long-range bombers, such as the Air Force's own pet project, the B-36, could fly anywhere at any time, drop devastating atomic weapons with precision on the enemy, and return home safely. Ships and armies, they argued, were necessary only to the extent that they could support the activities of the Air Force. For a country that had just lost hundreds of thousands of citizens in a bloody four-year struggle, it was a seductive theory: wars of the future would be won by invulnerable fleets of airplanes, high above the battlefield, with victory assured by American technical know-how. The Air Force argued convincingly that bombers should be flown by them alone. The Navy's job was to operate ships, they said, and atomic weapons had made ships obsolete.

For its part, the Army largely took the Air Force's side in the struggle. After all, the Air Force generals in the fight, men like Curtis LeMay, Hoyt Vandenberg, and George Kenney, had all been career Army officers until recently, and cultural ties between the two services were still strong. The Army, too, was historically accustomed to massive drawdowns after wars, and knew that it could just as rapidly be rebuilt, with the help of the still-active Selective Service System, the instant that shooting began anew. Additionally, the Air Force was more zealously going after the Navy's budget, fighting the Navy on every project it proposed that involved aviation. More so than the Army, the Navy and the Air Force both required years and millions of dollars to procure their expensive weapons platforms, and both believed with good reason that any budgetary gain of the other service was their loss.

Navy leaders were not being paranoid when they perceived that the nation's highest civilian leadership was biased against them, feelings that extended all the way to the White House. President Truman had been in the Missouri National Guard during World War I, and famously came to the conclusion that the Marines, a part of the Department of the Navy, were glory hounds who historically got more credit than they deserved. As president, Truman summed up his feelings about the Corps in a 1950 letter he wrote to Congressman Gordon McDonough of California: "For your information the Marine Corps is the Navy's police force and as long as I am President that is what it will remain. They have a propaganda machine that is almost equal to Stalin's." While Truman quickly backtracked when the letter was made public, no one in the Navy Department doubted his true feelings about the sea services. And while Truman's old grudge against the Marine Corps was worrisome, it was the disposition of Secretary of Defense Louis Johnson that kept the admirals awake at night.

Louis Johnson was only the second man to hold the position of secretary of defense. The job was created by the National Security Act of 1947, the same landmark legislation that created the Air Force as an independent entity, as well as the CIA and the National Security Council. Prior to 1947, the War Department, controlling the Army and the Army Air Forces, was a peer organization to the independent Department of the Navy. The 1947 act unified them all under the secretary of defense, and unification was something the Navy resisted, certain that it would make it subservient to the other services.

In part to placate the Navy, Truman named as first secretary of defense James Forrestal, a former secretary of the Navy who had initially opposed unification and was widely respected and trusted within the service. Unfortunately, Forrestal, his authority poorly defined and understood, found it impossible to quell the interservice rivalry that unification and shrinking budgets had inflamed. So stressful was the post that Forrestal resigned as secretary of defense after just eighteen months, citing exhaustion. He promptly checked himself into Bethesda Naval Hospital and seven weeks later jumped to his death from a sixteenth-floor window, on May 22, 1949.

Into this troubled position stepped Louis Johnson. In a role that required considerable tact, Johnson seemed almost to take pride in the number of enemies he could make. Historian William Manchester described him as "Truman's son-of-a-bitch." The urbane Secretary of State Dean Acheson, who despised Johnson, wrote that he was "mentally ill." General Harry Vaughan, a military aide to Truman, called Johnson "the only bull I know who carries his own china shop around with him." Even Truman, Johnson's patron, said of him later that he had a "pathological condition," that he "offended every member of the cabinet," and that he "never missed an opportunity to say mean things about my personal staff." Johnson was bald, big, and looked every inch like the ruthless, climbing politician that he was. He had helped found the American Legion after World War I, and was Franklin Roosevelt's assistant secretary of war. Ominously to those in the Navy Department, Johnson had been an aviation zealot even then, and was fired by Roosevelt in 1940 when France fell to the Nazis, revealing the degree to which the American military had deteriorated under his care (and that of his boss, Secretary of War Harry Hines Woodring, who was also fired). Johnson's political career was resurrected after he mobilized the American Legion, along with a considerable number of his own dollars, to support Truman in the 1948 race against Thomas Dewey. Truman appointed him secretary of defense in 1949, no doubt appreciating the talents of a man who could climb his way up the political ladder the old-fashioned way, by securing votes and calling in favors.

Johnson, for his entire career, was completely enamored with the idea that advances in aviation had made all other military endeavors obsolete. This was not conjecture on the part of paranoid Navy leaders, nor partisan criticisms of a man who had a legitimate mandate to slash defense budgets in a postwar world. Johnson really didn't like the Navy and he really loved the Air Force. He summarized his position neatly in a conversation with Admiral Richard Conolly in 1949:

> *Admiral, the Navy is on its way out. There's no reason for having a Navy and a Marine Corps. General [Omar] Bradley tells me that amphibious operations are a thing of the past. We'll never have any more amphibious operations. That does away with the Marine Corps. And the Air Force can do anything the Navy can nowadays, so that does away with the Navy.*

Johnson scrapped and sold off ships as fast as he could, ignoring the Navy's pleas to at least mothball some of them, in case his theory was wrong and the nation might someday again need destroyers and frigates. He did the same thing with the Army's tanks. Johnson was brazenly confident about his view of the future of war, and the evidence seemed to be piling up in his favor. When he ordered the cancellation of the *United States,* the Berlin Airlift had been going on for ten months, feeding and supplying the people of blockaded West Berlin entirely by air, humiliating the Soviets without a shot being fired and garnering massive American public support. The images of American candy bars floating down to the children of West Berlin on tiny parachutes were a symbol to many of how omnipotent airpower had become.

In addition to having visceral enemies in both the president and the secretary of defense, the Navy was also masterfully outplayed by the Air Force in the public relations battle between airpower and sea power. The new service portrayed itself as modern and forward thinking, and the Navy's admirals as fusty, gold-braided aristocrats from another century. Large ships, such as aircraft carriers, were vestiges of the pre-Hiroshima world, they argued: vulnerable, slow, and expensive. The Navy, unaccustomed to defending its very existence, sputtered at times that relying solely on strategic nuclear bombing was immoral, and at other times that it should be allowed to do the strategic bombing, too. *Time* magazine, a close ally of the Air Force's in the public debate, summed up the anti-supercarrier argument succinctly in its small story about the building of the *United States* before its cancellation: "The CVA-58 will probably carry about the equivalent of an Air Force bomber group, of which the Air Force has 16. One spread of torpedoes or a near-miss from an atomic bomb would put it out of action."

While trying to kill the supercarrier, the Air Force was promoting the airplane of the future: the B-36. The massive plane could fly 8,700 miles, more than 50 percent farther than the B-29, the most advanced bomber of World War II. The B-36 could carry eighty-six thousand pounds of bombs, twice the payload of the B-29, and five times the payload of the B-17. And unlike the theoretical advantages of a supercarrier, the B-36 already existed: the maiden flight of the plane took place on August 8, 1946. While the Navy could show drawings of the *United States,* the Air Force could fly its vision of the future, showing congressmen and their constituents alike what the shiny, hygienic future of warfare looked like.

With the Air Force generals smiling in the wings, Secretary of Defense Louis Johnson canceled the construction of the supercarrier without consulting anyone in the Navy. An enraged Secretary of the Navy John Sullivan resigned in protest. A group of desperate Navy officials drafted an anonymous document that implied that Louis Johnson and the secretary of the Air Force, Stuart Symington, were benefiting financially from the B-36 program. A congressional investigation was hastily convened.

While the charges of corruption were baseless, the hearings did give the Navy a chance to plea publicly for its survival. Starting on October 6, 1949, a string of Navy officers testified, including well-known heroes from World War II: Admirals Ernest King, Chester Nimitz, William "Bull" Halsey, and Raymond Spruance all spoke up, as did Marine Corps Commandant General Alexander Vandegrift. They testified about strategic bombing, the Air Force, and the B-36 bomber. The USS *United States* wasn't vulnerable, the Navy argued—the B-36 was.

Admiral Arthur Radford was the Navy's most compelling witness. A naval aviator since 1921, Radford had, among other duties, commanded Carrier Task Force 6 during the war, directing the naval aviation that helped win the battles at Iwo Jima and Okinawa. Radford was a man who could speak convincingly about the advantages of a strong naval air wing. He was also dashingly handsome, a good speaker, and believed fervently everything he said. Radford described the B-36 in much the same way the Air Force described the supercarrier: "slow, expensive, very vulnerable . . ." He attacked the Air

Force's claims about the accuracy of the bomber. "The B-36 cannot hit precision targets from very high altitudes under battle conditions."

If Radford was the Navy's most convincing witness at the hearings, the most anticipated was the chief of naval operations, Admiral Louis Denfeld. No one was quite sure what he would say. For him to support the views of his aviators would be to take on his colleagues within the Joint Chiefs of Staff. In an environment where allegiances and loyalties were being scrutinized, Denfeld was not a naval aviator—he had spent most of his seagoing career on destroyers. In the end, Denfeld decided to commit professional suicide rather than betray his fellow naval officers.

"Why do we need a strong Navy when any potential enemy has no Navy to fight? I read this in the press, but, what is more disturbing, I hear it repeatedly in the councils of the Department of Defense." With those words, Denfeld began a detailed defense of the Navy, and a repudiation of the beliefs and the abilities of his boss, Louis Johnson, the secretary of defense. "There is a steady campaign," he told Congress, "to relegate the Navy to a convoy and antisubmarine service . . . this campaign results from a misunderstanding of the functions and capabilities of navies and from the erroneous principle of the self-sufficiency of air power." Denfeld would be fired within days.

General Omar Bradley, war hero and brand-new chairman of the Joint Chiefs, provided the strongest statement of pure interservice vitriol to close the hearings. To impugn Denfeld's testimony, he first recounted his own impeccable war record, as commanding general of the Twelfth Army Group, at 900,000 men the largest field command in the history of the U.S. military. As for Denfeld, the general stated "I was not associated with Admiral Denfeld during the war, and am not familiar with his experiences." Denfeld, as the general well knew, had safely spent most of the war in Washington as assistant chief in the Bureau of Navigation. Known by the public as a polite and calm counterweight to the bellicose George S. Patton in Europe, the strength of Bradley's emotion in the congressional hearings shocked everyone. He saw the admirals' testimony as not just obnoxious but disloyal, an attack on the principle of civilian control of the military. His voice trembling with anger, Bradley stunned the hearing room by calling the Navy admirals "fancy dans who won't hit the line with all they have on every play, unless they can call the signals."

History would call the entire episode the Revolt of the Admirals, and like most rebellions, this one was doomed. The USS *United States* remained canceled, and the primacy of the Air Force was affirmed as the backbone of the nation's Cold War military doctrine. Louis Johnson and the Air Force had won the day. The leadership of the Navy, cowed by their defeat, moped on the sidelines, hoping they might someday get another chance to prove that navies were somehow still relevant in the atomic age.

So the blueprints for the USS *United States* were rolled up and shelved, referred to by only those naval aviation fanatics who refused to let die the dream of the supercarrier. Few aviators probably bothered to look at the plans for the engine room, the least loved and most important compartment on any warship. The engines of the *United States* would need to generate an incredible 280,000 shaft horsepower to propel the ship at the required 33 knots. The engine room would use the latest postwar developments of high pressure (1,200 psi), high temperature (950° F) steam, generated by eight oil-burning boilers in four engine rooms.

But while unusually big, and perhaps incrementally hotter and at higher pressure, there was nothing revolutionary about the engine room of the *United States*. In burning fossil fuel to boil water and create steam, the *United States* was propelling herself in fundamentally the same way that Robert Fulton had propelled his ship down the

Hudson in 1807. (Interestingly, the Air Force's B-36 was also propelled in a way that would soon be seen as quaintly old-fashioned—it was the largest airplane ever built with piston engines, the last of its kind at the dawn of the jet era.) The USS *United States*, the Navy had believed, was a carrier for the atomic age, because she could carry planes large enough to be armed with atomic bombs—that's why they had proudly added the "A" to her hull number. Only a few thought there might be some other practical use for the mysteries of the atom, a use that would truly revolutionize the Navy, propel it back to the vanguard of the nation's war plans, and create an entirely new source of power in the process.

The man who would rescue the Navy was Hyman George Rickover, a five-foot, five-inch-tall captain who never served in combat and wasn't even eligible to command a ship—as an engineering duty officer he was limited to vital but inglorious jobs such as managing shipyards and supply depots. Laboring deep within the bowels of an obscure bureaucracy, this captain was convinced that by the sheer power of hard work, discipline, and intellect, he could create a use for the atom other than nuclear weapons. In an incredibly short period of time, he would almost single-handedly take this idea from the most skeletal of theories to the fulfillment of an ancient naval dream. The swiftness of his success surprised everyone, on both sides of the Iron Curtain. When he burst on the scene, however, he was not greeted as a savior by the admirals who had survived the dark days of 1949. Instead, they did everything they could to throw him out of the Navy.

THE CADRE

The technical investigation into the SL-1 explosion would involve hundreds of engineers over a period of years, men and women in the employ of a dozen different entities, including the National Reactor Testing Station, the Atomic Energy Commission, General Electric, and Combustion Engineering. The investigation would assume the character of many a nuclear enterprise as it became a slightly bloated, self-sustaining bureaucracy, an interesting occupation for mostly anonymous men who churned out thousands of pages of mind-numbing detail about what had gone wrong with the machinery of SL-1.

The investigation into the personal lives of the three men killed at SL-1 was much less expansive, and would for the most part bear the signature of a single man: Leo Miazga. Miazga was a tough investigator who came out of the mountains of eastern Pennsylvania and won two Bronze Stars in World War II fighting in Europe and Africa. As special investigator for the Atomic Energy Commission, Miazga would author two scant memos about the personal lives of Legg and Byrnes, a total of thirty-one double-spaced pages that represent the only contemporary investigation of the personal lives of the three men who died that night.

• • •

Even now, one of the first things people who knew Richard Legg say about him is that he was short. The autopsy report would measure him at five-and-a-half feet tall. Self-consciousness about his stature may have caused the characteristics people usually mention next when asked to describe Richard Legg: a certain belligerence, a drive to constantly physically assert himself, a vanity about the muscles he worked to build, flex, and use in roughhousing that always bordered on the aggressive. The secret memo written by Miazga in 1962 actually used the words "small man complex."

Legg was a Navy man at an Army nuclear plant, a proud member of the Navy's famous construction battalion, a Seabee, with the requisite bumblebee tattoo on his right shoulder, and a hula girl tattooed on his left arm for good measure. The Army was required to allow a certain number of Navy men into its small reactor program because of the sea service's interest in portable land-based reactors. Rumor had it that the Army drew all its Navy men from Seabees in order to keep Rickover's meddling hands off the Army program, something they couldn't do if it were staffed with Rickover's regular "nukes," the elite, hand-chosen nuclear-trained men from the submarine force.

Legg grew up in Roscommon, Michigan, a tiny, rural town in the central part of the state where his father owned a sawmill. He joined the Navy in Grayling, Michigan, the nearest town large enough to host a recruiter, shortly after graduating from high school in 1952. After two years in the Seabees, Legg read in a brochure that the Army was recruiting men from other services into its nuclear program. Legg reached the same conclusion the Army had, that the America of the future was atomic-powered, and he signed up. He arrived at Fort Belvoir, Virginia, to begin his training in 1958.

During the eight-month course at Fort Belvoir, Legg, like the rest of the men, was evaluated for his overall fitness, mental and physical, for isolated duty. Isolated, because the Army justified its nuclear program largely by citing the power requirements for remote, Arctic bases. Having successfully navigated all qualifications placed in front of him, Legg was ordered to SL-1 in Idaho to complete his training in the fall of 1959. He was twenty-five years old.

Legg did not come to Idaho with a wife, but he soon found one, among the modest Mormon families of the area. She was the nineteen-year-old Judith Cole, who lived with her devout parents on 265 South Capital Avenue in Idaho Falls. She worked as a stenographer on the site. While her parents were perhaps distressed at the seven-year age difference between them, and that Legg was not Mormon, they probably took solace in the fact that she had married a man in the nuclear program. Such a marriage was seen as upward mobility for the young women of the area. At the time of SL-1's explosion, she was eight months pregnant with their first child.

In Idaho, Legg progressed through his training, eventually qualifying as shift supervisor and chief operator in September 1960. Legg's professional progress was not unblemished, however, and he demonstrated some of the same problems with authority that would most often be attributed by future investigators to Jack Byrnes. Soon after Legg qualified as shift supervisor, the SL-1 plant superintendent caught him with his feet on the SL-1 instrument panel. Unfazed, Legg at first refused to remove them. He'd also been caught sleeping in his car while on duty, and showed up for work drunk a month before the explosion. Legg would set off alarms intentionally to startle his crewmates, and on one occasion he actually shut off a cooler, which caused some control instruments to overheat.

The most serious disciplinary incident in Legg's career took place in December 1960, when he falsified records to show that a friend was on watch inside the plant when he was actually absent. Unfortunately for Legg, his friend was spotted by a supervisor in downtown Idaho Falls at that same time. As a result, the colleague was moved to a different crew so that he would no longer be supervised by Legg. His replacement on Legg's crew was Jack Byrnes.

Another consequence of the incident was that Legg was reprimanded by Sergeant Paul Conlon during the staff Christmas party at the Rogers Hotel on December 23, 1960. When confronted, Legg actually puffed out his chest and challenged his superior to a fistfight. Sergeant Conlon declined, and told Legg that he expected to see him to discuss the matter in the SL-1 office the next time they were on-site at the same time. The day of the scheduled meeting: January 4, 1961.

While Richard Legg was known for his short stature and his attempts to compensate for it, Richard McKinley was actually the smallest man on the crew. He was the same height as Legg, but weighed just 115 pounds compared to Legg's stocky 160. McKinley had been in Idaho just three weeks, but under different circumstances might have been the ideal moderator between his two volatile crewmates. Born in Union City, Indiana, in 1933, he was the oldest of the three men at twenty-seven. He was a veteran of the Korean War. He was happily married to Caroline and had two small children, ages three and nine months. He was also a career military man, having served four years in the Air Force, one and a half years in the reserves, and another four years in the Army, four years that ended with his tour in Idaho. With a few more weeks, perhaps the mature, midwestern family man could have helped avert whatever developed inside SL-1 on January 3, 1961. As it was, he was too inexperienced either to prevent the disaster, or to cause it.

Jack Byrnes was bigger than Legg, at five feet, ten inches and 175 pounds, something they both surely noticed in the hypercompetitive, masculine atmosphere of the Army's nuclear power program.

Like Legg, Byrnes was married. His wife was the former Arlene Casier, his hometown sweetheart from Utica, New York. They wed before he left New York for the Army, after forging his birth certificate to make him one year older than his seventeen years, in 1956. They had to grow up quickly—by the time he was nineteen, Jack Byrnes had a wife, a child (John Byrnes IV, born in 1958), and an enlistment contract with the U.S. Army.

Byrnes served a tour in Newfoundland, Canada, where he must have found something about the frigid climate that agreed with him. He volunteered for the Army's nuclear program in 1958, and was ordered to Fort Belvoir, where one of his classmates would be Richard Legg. After successfully completing the training at Fort Belvoir, Byrnes received orders to report to SL-1 in Idaho.

Like Legg, Byrnes seemed not to have outgrown some of his youthful tendencies by the time he arrived in Idaho. He complained extravagantly at work, and threw tools in frustration at any perceived slight. He once refused to conduct a check of the plant as ordered by Sergeant Herbert Kappel, his chief operator before he was transferred to Legg's crew. Kappel offered to take Byrnes outside and teach him a lesson in respect, an offer Byrnes declined, and an offer that prompted him to comply reluctantly with Kappel's order. In the United States Army of 1960, Kappel's training techniques were sanctioned by the chain of command. Before challenging Byrnes to a fight, Kappel had discussed the idea and received approval from Sergeant Richard Lewis, the plant superintendent, the same superintendent who had ordered Legg to remove his feet from the instrument console. Kappel told Miazga later that "this approach was selected as a means of assuring that he take a more serious view of his job."

While his interactions with his fellow soldiers were rocky, his stormy relationship with his young wife was at the center of most of his worst moments. Byrnes's colleagues, when interviewed by investigators after the explosion, almost universally agreed with the statement given by Jim Bleak, one of Byrnes's coworkers at Kelly's Texaco in Idaho Falls where he worked part-time. Bleak said that Byrnes "preferred the company of male friends to that of Mrs. Byrnes" and that "the only occasions that Mr. Byrnes displayed temper was during conversations with his wife."

Byrnes had learned to ski while growing up in upstate New York, and he soon found that his new home in Idaho was a skier's paradise, with mountains in every direction. In order to ski for free, he volunteered for the ski patrol in the Pine Basin Ski Area. He would spend long hours on the slopes, almost always without his increasingly frustrated wife, who was at home raising their young son on a tiny budget, without the benefit of the many diversions that Jack Byrnes had created for himself in Idaho.

As a result, when he was home, they fought constantly. Robert Matlock, an engineer and onetime neighbor of Byrnes, told Miazga with an engineer's precision that the couple "had serious arguments on an average of one a week and less serious ones in between." Some of their fights became public spectacles, with Arlene throwing Jack's clothes onto the lawn as she screamed. The military community was sensitive about such antics in the small town that hosted them; Sergeant Gordon Stolla's wife counseled Arlene that "civilians generally class military personnel as a group and if one of a group gets a bad name the name is applied to the rest of the group." Mrs. Stolla told Miazga that her advice "made no impression on Mrs. Byrnes."

Mrs. Stolla was the only witness interviewed who seemed to blame Arlene Byrnes almost entirely for the problems in that marriage, and she also spoke very favorably of Jack Byrnes. She was impressed the first time she ever spoke to Byrnes, by his "attitude and stated goals," and thought his voice sounded like that of her little brother's. He was, in her words, "kind, considerate, and polite," "intelligent and eager to learn."

Arlene Byrnes, on the other hand, seemed to annoy Mrs. Stolla with her every act, and it was Arlene's failure to uphold her high standards for proper Army wife behavior that seemed to most offend her. Miazga wrote of Mrs. Stolla: "She said that Mrs. Byrnes wanted to have a nice apartment and good furnishings but made no real attempt to preserve and maintain her furniture or personal effects and as a result was actually losing instead of gaining the things she desired."

Even Arlene Byrnes's performance as a widow was criticized by Mrs. Stolla. She, again alone among those interviewed, described Arlene after Jack's death as "completely mercenary" and said that she "dwelled at length as to how much money she would get as a result of the loss of her husband and how soon it would be given to her."

Perhaps because her opinions were so strong and not echoed by others, Leo Miazga thought it prudent to look into the background of Mrs. Stolla as well. He checked with Captain R. L. Morgan, ranking officer for the whole site, who told Miazga knowingly that "Mrs. Stolla is a frank and somewhat outspoken person." He added, "for this reason some people tend to dislike her."

During the holiday shutdown of the reactor, Jack and Arlene actually spent some time together, perhaps in an attempt to heal their marriage, or perhaps at the insistence of Arlene. On December 23, they attended the eventful staff Christmas party at the Rogers Hotel together, the same party where Legg challenged his supervisor to a fistfight. They attended a New Year's party together on December 31. Perhaps that went well, because the next day Byrnes took his wife with him on what was his most treasured solitary pastime: skiing.

They returned home from the slopes at 5:00 PM, but apparently their time together had not repaired the underlying damage to their marriage. They began a violent argument, intense even by their standards. At one point, Arlene went to Robert Matlock's apartment next door to call the police. While she was on the phone, Byrnes left. For the next two nights he would sleep at the apartments of friends. Byrnes returned home on the morning of January 3 to look in the mailbox for his Army paycheck and was enraged when he found that Arlene had taken it. He didn't have time to pursue Arlene, the check, or another fight. The holiday break was over, and he had been assigned to a new crew, Legg's crew. His shift began at 4:00 PM.

fashion, the congressman nominated the plucky delivery boy from his district for the Naval Academy, where Rickover graduated in 1922 as a good but not spectacular student, ranking 107th out of 540 midshipmen.

After graduating, Ensign Rickover first served on destroyers: the USS *Percival*, followed by the USS *La Vallette*. In 1925, he reported to the battleship USS *Nevada*, which would become famous for heroically getting under way while bombs exploded around her at Pearl Harbor on December 7, 1941. Nothing that glamorous occurred during Rickover's time on board. From the start, however, Rickover showed a consuming desire to learn every detail of his ship, with an intensity that alienated some. He studied ships' drawings and technical manuals day and night, and made few friends.

From the *Nevada*, Rickover went to Columbia University, where he obtained a master's degree in electrical engineering in 1929. While at Columbia, Rickover met Ruth Dorothy Masters, an international law student whom he would soon marry. The ceremony was conducted by an Episcopal priest, a fact that estranged Rickover from his parents for decades. After Columbia, Rickover traveled to New London, Connecticut, and reported to the submarine *S-48*, on June 21, 1930. It was the beginning of his longest tour on any one vessel, and his only tour aboard a submarine, the type of ship he would soon revolutionize.

Submarines had always been seen as short-range vessels of coastal defense, a far cry from the majestic battleships and the "Command of the Sea" philosophy advocated by Alfred Thayer Mahan and his fellow prophets of a large-ship, open-ocean navy. In mission and construction, the *S-48* was a direct descendant of the USS *Holland*, the Navy's first submarine, purchased just twenty years before the keel of *S-48* was laid. The *Holland* was built by and named for John Holland, an Irish schoolteacher and patriot who had built his first submarines for the express purpose of killing Englishmen and sinking English ships. Although he arrived in the United States in 1873, Holland's first submarines were funded by the rebel group the Irish Republican Brotherhood—one of his early attempts was christened the *Fenian Ram*. As his designs became more sophisticated, Holland tried to catch the attention of the U.S. Navy, actually winning a design competition for a submarine in 1888. But Holland soon grew tired of the straitjacketing influence of Navy bureaucracy and conservatism. Holland built a privately financed submarine of his own design in 1898, the one he thought the Navy should be buying for themselves. Two years later, after an exhaustive evaluation, the Navy agreed and purchased the vessel for $150,000.

The U.S. Navy had dabbled with various underwater propulsion schemes before. The *Turtle,* powered by a single pedaling crewman, attacked a British warship unsuccessfully during the American Revolution. The first warship ever sunk by a submarine was the Union's *Housatonic,* sunk by the Confederate *Hunley* near Charleston in 1864. Despite these earlier efforts, the U.S. submarine service still celebrates the sale date of the *Holland* as its birthday: April 11, 1900. Thus the service was founded by an Irish rebel who had to step outside the Navy bureaucracy to get his ship built, an interesting foreshadowing of Rickover's career.

Lieutenant Rickover's submarine would have been instantly recognizable to John Holland. The *S-48* was essentially a diesel-powered surface ship that could, for short periods of time, submerge and operate on an electric battery. The top speed for the *S-48* on the surface was 14.2 knots. Her underwater endurance was limited to twenty hours at 5 knots. Perhaps the most salient characteristic shared by all early submarines, including the *S-48,* was the peril of serving inside one. The *Hunley* had sunk and killed her entire crew of eight during her attack on the *Houstonic*—prior to that she had killed her namesake, H. L. Hunley. Submarine safety had advanced little in the intervening years. Rickover's *S-48,* prior to his arrival, had flooded and sunk in shallow water during her sea trials in 1921—the crew and the chagrined builder's representatives escaped through a torpedo tube. She sunk again off New England during a heavy storm in January 1925. During Rickover's time on board, in 1930, a fire broke out in the ship's battery well while on the surface—a perennial danger in boats that depended on the stored chemical energy of the volatile, acid-filled cells. During that fire, with the rest of the crew taking refuge topside, the intrepid Lieutenant Rickover donned a gas mask and entered the battery well to verify the fire had burned itself out. So accident-prone was the *S-48,* in fact, that she had the unusual distinction of being decommissioned three separate times before being stricken permanently from the naval register in 1945.

Rickover was promoted to executive officer of the *S-48*, the second in command, and completed the qualification process to command the boat. Before that could happen, however, following the Navy's traditional route of running its officers through a wide variety of ships and billets, Rickover was ordered to the battleship *New Mexico* as assistant engineering officer in April 1935. Aboard the *New Mexico,* Rickover continued the relentless—some would say ruthless—pursuit of engineering excellence. The *New Mexico* was one of fifteen battleships that competed for overall fuel efficiency, a crucial concern during a time of austere interwar defense budgets. When Lieutenant Rickover reported on board, the *New Mexico* was squarely in the middle of the pack, ranked eighth. By dimming lights, turning down the heat, and timing hot showers with a stopwatch, Rickover soon made the *New Mexico* the most fuel-efficient ship in the group, a position she would maintain for three straight years. While Rickover's accomplishments were greatly appreciated by his superiors, the men whose showers he timed were less grateful. And Rickover's intensity was not leavened by the kind of jocular backslapping that has always been an unwritten requirement of Navy wardrooms. He preferred the company of books to his fellow junior officers, and openly disdained the parties and social calls that were an absolute obligation of an ambitious naval lieutenant, then and now.

His efficiency, dedication, and unparalleled ability to get things done trumped his social liabilities for the moment, and Rickover was promoted to lieutenant commander in 1937. He transferred to the USS *Finch*, a minesweeper off the coast of China, the one and only ship Rickover would command during his career. He would serve in that capacity for slightly less than three months, after which he would begin his new professional life, as an engineering duty officer (EDO).

In later years, when Rickover was firmly entrenched, his status as an EDO would often be commented on disparagingly by line officers. It was an old prejudice, those who fancied themselves the Navy's warrior class looking down on specialists of any kind. The EDO billet had been created reluctantly by the Navy in 1916 in recognition of the increasing engineering complexity of steam-powered warships. They needed men who could design, procure, and maintain the machinery that had replaced sail power. In a proud seagoing service, EDOs like Rickover were for the most part confined to land. In a profession with no higher aspiration than to captain a ship, EDOs were ineligible to command.

In his case, Rickover's small physique sometimes furthered the impression that he was somehow not a normal naval officer. Elmo Zumwalt, a future chief of naval operations, would describe him as "gnomelike." The first chairman of the Atomic Energy Commission (AEC), David Lilienthal, wrote in his journals of Rickover, "There is something exceptional about his face. It is a small face, almost as if one were looking at him through the wrong end of a telescope—that kind of smallness." Combined with his unapologetic passion for the minutiae of ship design, Rickover looked at first glance like a stereotypical, slide-rule-carrying engineer. In reality, before Rickover even began his EDO career, he had been a line officer for over fifteen years, had served on some seven different warships, and commanded one. It was a seagoing resume that many of his "warrior" critics couldn't match.

Rickover would serve as an EDO for the duration of World War II, in a string of vital but obscure billets. Most of the war he spent as the head of the electrical section in the Bureau of Ships, in Washington, D.C., where he would begin his lifelong practice of terrorizing government vendors with midnight phone calls and hastily called weekend meetings. He also developed a number of curious habits that troubled many of his fellow officers. When a man reported to Rickover, the captain placed him in a position that had everything to do with his qualifications and experience, and little to do with his rank. In many cases, shocked commissioned officers found themselves reporting to civilians because they lacked expertise, protocol be damned. Sometimes—horrors—they even reported to enlisted men. This same disdain for Navy custom would soon manifest itself in an even more visible way. Rickover began to show up for work, and even highly public events, in a civilian suit rather than his naval uniform. Decades later, whenever one of Rickover's military critics would begin a reasoned critique of his methods, they would often finish seething about the sight of Rickover in civilian clothes, something they often viewed as a sneering attack on the whole military culture. The Web site of one submariner, a career enlisted man who admitted his personal interactions with Rickover were "one step less than minimal," shows how the image of Rickover in a suit still grates many of those who spent their lives in uniform, even years after the Rickover era:

> *If you are an Admiral you should look like one. Admirals have great looking blue uniforms with lots of gold on the hat and sleeves. They also wear rows of ribbons and medals. Rickover spent most of his naval career dressed in civilian clothes. If he wanted to be a civilian he should have resigned from the Navy*

and became one. We were proud of our uniform, he should have felt the same.

In a service that sanctified the gold braid, ribbons, and brass buttons of rank, Rickover dismissed it all, and one senses that this, above all else, is what really infuriated many of his critics in the military. One also senses that this is exactly why Rickover kept on doing it.

Near the end of the war, Rickover was given the task of whipping into shape an underperforming supply depot in Mechanicsburg, Pennsylvania, a task he did with characteristic zeal and efficiency, leaving a trail of enemies in his wake, all those he felt were not giving the Navy their full effort and therefore deserved none of his very limited supply of mercy. From Mechanicsburg he went to a supply depot in Okinawa, the closest he would get to a combat zone during the war. His duty station was demolished by Typhoon Louise in 1945 and abandoned by the Navy as not worth repairing. As the war ended, Rickover was given perhaps the ultimate dreary postwar duty for an EDO, that of supervising the mothballing of Navy ships as they came off the battle lines and went into long-term storage in San Francisco Bay.

As Rickover probed the bilges of abandoned ships looking for "painted over banana peels," or any other sign of dereliction, he might have contemplated his past two decades in (and out of) uniform. He had served twenty-three years, commanded a ship, and achieved the rank of captain. He had, in other words, completed a successful naval career and was eligible, by any measure, for retirement. Rickover, however, was about to embark on a new career within the Navy, one ideally suited for his feisty mix of engineering skill, experience with civilian industry, and relentlessness. In May 1946, Rickover was one of five naval officers sent to Oak Ridge, Tennessee, to participate in a small study to determine whether nuclear fission might power a submarine.

• • •

It is hard to comprehend now just how outlandish this notion was. The neutron had been discovered just fourteen years before, by James Chadwick in his Cambridge, England, laboratory. The first nuclear bomb had exploded just ten *months* before, on July 16, 1945, in Alamogordo, New Mexico. The very few working atomic reactors in the world were designed to create plutonium for nuclear bombs; no one had yet tried to generate usable quantities of power with them. To design a reactor that would be dependable, battle-hardened, or even fit inside a submarine seemed generations away. But the promise of virtually unlimited power made the pursuit irresistible.

Nuclear fission results when the nucleus of an atom, its tight bundle of neutrons and protons, splits. This results spontaneously, at times, in some unstable elements, or when the nucleus is struck by a neutron. Several important things result from this split. It produces entirely new elements, the fragments from the original nucleus. These elements can have properties that inhibit fission, like xenon, or are nuclear fuel themselves, like plutonium. Fission also produces gamma rays, a form of radiation essentially identical to x-rays, and neutrons, which can, if enough remain in the vicinity of enough fuel, cause more fissions and sustain a chain reaction. Finally, and most importantly to Rickover and his colleagues at Oak Ridge, the splitting of the atomic nucleus releases enormous amounts of energy.

The energy released is the "binding energy" that previously held the nucleus together. This release of energy results in a small reduction in the mass of all the constituent parts, mass that is converted into pure energy in accordance with Einstein's famous equation, $E = mc^2$. E in the equation is energy, and it is proportional to the mass of an object multiplied by the square of c, the speed of light. The conversion of matter to energy in man-made nuclear devices is far from perfect: the destruction of Hiroshima resulted from the conversion of about one gram of matter into energy, roughly the weight of a dollar bill. That small amount of mass, when converted to energy, resulted in an explosion roughly equal in force to 15,000 tons of dynamite. Nuclear proponents have long struggled to put this enormous potential energy into understandable terms. The fission of one pound of uranium equals 6,000 barrels of oil, equals 22,000 pounds of coal, equals 14,000 pounds of liquid natural gas. In a nuclear bomb, that energy is released in an instant, while in a nuclear power plant it is released slowly, over months or years.

Fission is more likely to occur in elements with heavy, unstable nuclei, and uranium is the heaviest element to occur in nature in any appreciable quantity. This makes uranium the link between the natural world and the sometimes through-the-looking-glass world of nuclear physics, a world of "transuranic" elements that exist only in particle accelerators and reactors for milliseconds at a time, substances named for the main characters in the early dramas of atomic physics: einsteinium, curium, fermium, seaborgium, and so on. Plutonium is another example, and an important one, as it can also be fuel for a nuclear reactor, and is superior to uranium as the fuel for a nuclear bomb. All these elements must be manufactured with a nuclear reaction, a reaction that is ultimately powered by the fission of uranium. To put it another way, there are uranium mines all over the world. There is no such thing as a plutonium mine.

In nature, 99.2798 percent of uranium is the isotope uranium 238, meaning that it has 92 protons and 146 neutrons in its nucleus. Uranium 235—92 protons with 143 neutrons—makes up a scant .72 percent of natural uranium. For reasons that defy explanation to all but physics PhDs, U-235 is much more susceptible to fission than the more prevalent U-238. While it is possible to build a chain reaction with naturally occurring uranium—Enrico Fermi did it beneath the stands of Alonzo Stagg Field at the University of Chicago—it is extraordinarily difficult, takes up vast amounts of space, and generates little power in return for the trouble. The trick for efficient fission is to concentrate the U-235. Uranium that contains more than its natural .72 percent of U-235 is said to be "enriched." Since U-235 and U-238 are chemically identical, separating one from the other is an enormous challenge, and was, in fact, one of the primary challenges the Manhattan Project team faced down as they built the first atomic bomb. They eventually invented a way to gasify uranium by combining it with fluoride, a method still in use today. The resulting gas, uranium hexafluoride, is processed through osmotic filters that allow the different isotopes of uranium to be separated.

Uranium is not that rare. It is more common than gold, silver, mercury, or tungsten, and roughly as plentiful as tin. Uranium in the form of one of its natural oxides, U_3O_8, currently sells on the world market for about ninety dollars per pound. Canada, Australia, and Niger are the three biggest producers.

In nature, uranium occurs in several different forms, the most famous of which is the mineral "pitchblende." The eighteenth-century German miners who first identified pitchblende in the silver mines of Saxony portentously named the black rock after the words *pech* for "bad luck" and *blende*, for "mineral." Within pitchblende a self-taught German chemist named Martin Klaproth first identified the element uranium in 1789. He modestly declined to name it after himself, but instead named it for Uranus, the planet discovered just eight years earlier. It was in pitchblende that Marie Curie discovered another new element, radium, in 1898, a discovery that would make her the first woman to win a Nobel Prize in 1903. Before the atomic age, uranium was a pedestrian element of little use to society, used primarily as an additive to vivid red glazes for pottery. It was a property of uranium still being exploited in the production of "Fiestaware" plates until the 1940s, when the U.S. government seized control of the uranium supply for reasons the potters couldn't begin to fathom.

• • •

Besides the prospect of virtually unlimited power, another key attraction of atomic energy to the Navy was this: nuclear fission requires no oxygen. A submarine powered by the atom could stay underwater, invisible, for months at a time, freed from its snorkel mast and detached from the atmosphere above, her patrols limited only by the duration of the crew. It was truly the stuff of science fiction, a nearly invincible ship, like the *Nautilus* in Jules Verne's *20,000 Leagues Under the Sea*. As Rickover well knew from his experiences aboard the *S-48*, nuclear propulsion would change the very nature of submarining, and perhaps the nature of warfare. Advocates referred to the ideal as the "true submarine," and that singular language speaks to its dreamlike hold over seagoing men.

Rickover, with no background in nuclear physics whatsoever, tackled the challenge with vigor, solving engineering problems and stepping on those who got in his way with equal enthusiasm. Part of Rickover's genius was in approaching the problem of the nuclear submarine from the start as an engineering challenge, rather than a scientific one. The science had already been worked out, he argued, thanks to the Manhattan Project. Oak Ridge, for example, where Rickover first studied nuclear power, was where the government had perfected a process for enriching uranium. Now they had the difficult but entirely possible task of assembling the pieces together in a way that would power a submarine: a job for engineers. Rickover quickly marshaled the forces of two giants of U.S. industry to build competing models for submarine reactors: General Electric and Westinghouse. General Electric would build a reactor cooled with liquid sodium, while Westinghouse built one cooled with pressurized water.

General Electric had been associated with nuclear power from the beginning. In 1946, as the Manhattan Project was winding down,

DuPont announced to a frantic federal government that it was no longer interested in running the plutonium production facility at Hanford, Washington. GE was one of few American companies with the industrial wherewithal to run such an operation, but it was uninterested in tying its future to the business of manufacturing nuclear weapons. To sweeten the pot, just weeks before the Atomic Energy Act took effect, General Leslie Groves agreed to give GE the not-yet-born nuclear power facility that would become Knolls Atomic Power Laboratory, or KAPL, near Schenectady, New York. In addition, the May 15, 1946, contract completely indemnified GE against anything that might happen at either facility, "in view of the unusual and unpredictable hazards involved in carrying out the work under this contract." Ten weeks later, the civilians took control. The AEC's first commissioner, David Lilienthal, reviewed the contract with his team and was horrified to see what a great deal GE had received: "We spent most time on a contract with General Electric for the operation of Hanford and the operation of an expensive laboratory at Schenectady which the contract provides the Government will pay for. When I first heard of this, I didn't like it; didn't like it at all." Nonetheless, Lilienthal recognized that the contract had been signed in good faith by the U.S. government and felt obliged to honor it.

Rickover was also negotiating unusual terms with the bureaucracy. In a masterstroke, he maneuvered himself into key positions both within the Navy's Bureau of Ships and in the civilian Atomic Energy Commission, garnering two titles only a bureaucrat could love. He was simultaneously the assistant chief of the Bureau of Ships for Nuclear Propulsion within the Navy, and chief, Naval Reactors Branch, Division of Reactor Development within the AEC. When it was required, he would write himself letters and immediately respond, taking full advantage of his ability to command two different powerful letterheads. Rickover had by this time fully developed a philosophy that led to one of his most enduring quotations: *It's better to sin against God than against the bureaucracy. God will forgive you, the bureaucracy won't.*

In 1949, his task acquired new urgency, both for the Navy and the nation. In April of that year the USS *United States* was canceled by the Navy-hating Secretary of Defense Louis Johnson. Some saw Rickover's submarine, if the thing was possible, as a way to regain a measure of prestige for the Navy. Four months later, on August 29, 1949, the Soviets exploded their first atomic bomb, surprising Johnson so much that for days afterward he chose not to believe it, until a committee headed by Manhattan Project hero Robert Oppenheimer told him emphatically that it was true. By then, although not a single bolt had been turned on the new submarine, Rickover confidently declared his nuclear submarine would get under way on January 1, 1955.

Money was difficult to come by for a major military project in a country at peace. Rickover estimated his submarine would cost the taxpayers $40 million. More so than money, one of the main factors limiting Rickover's progress was the intense competition for the nation's very limited supply of enriched uranium. In 1949, the nation's entire U-235 supply could be measured in pounds. The overwhelming priority of the government was to build more atomic weapons—the nation had used up its entire nuclear stockpile with the two weapons dropped on Japan in 1945. As those stockpiles grew, however, Rickover found that he wasn't the only one interested in building a nuclear reactor for propulsion.

The Air Force's pursuit of an atomic airplane was in some ways more advanced than the atomic submarine in those early days. The same month the Navy had detailed Rickover to his exploratory mission at Oak Ridge, May 1946, the Army Air Forces had awarded a contract to the Fairchild Engine and Airplane Corporation to determine the feasibility of a manned, nuclear-powered airplane. The Air Force had visions of atomic-powered strategic bombers that could stay in the skies for weeks at a time, circling the world until called upon to pound the commies into oblivion. Just as for Rickover, it was a highly speculative venture at a time when power-generating nuclear reactors of any kind had yet to be invented, much less one that could fit inside an airframe. In 1946, however, the Air Force's vision of its atomic future was not any more fanciful than the Navy's. And the Air Force had an added advantage—many powerful people believed the argument that all future wars would be won by strategic bombing, not by something as old-fashioned as a submarine.

Louis Johnson, the Air Force, and their disciples in the Truman administration were dealt a stunning blow on June 25, 1950, when the communist North Koreans began pouring over the 38th Parallel. The United States found itself in a shooting war again, and discovered to its horror that it had very few guns left with which to shoot. In a short period of time, Soviet conventional weapons, both in quantity and quality, had surpassed the weapons of the United States. The Air Force generals, who had devoted almost all their resources to the B-36 strategic bomber, began receiving alarming reports from their neglected fighter pilots about the superior performance of the Soviet-made MiG-15 fighter jets. As for the rest of the country's conventional arms, Louis Johnson had let the nation's tanks, ships, and guns rust away, since he was confident that such weapons were obsolete in the Atomic Age. With the fundamental theorem of his tenure proven tragically wrong on the battlefields of Korea, and less than eighteen months after brashly canceling the construction of the USS *United States*, Louis Johnson was fired by President Truman on September 15, 1950. He spent his last moments as secretary of defense blubbering and begging for his job in the Oval Office as a mortified Truman ordered him to sign his resignation letter.

The firing of Johnson did not completely deter those who thought the United States could (and should) win its wars with nuclear weapons. As the Chinese joined the fight in Korea, pushing the beleaguered

American forces farther and farther back, General Douglas MacArthur developed a detailed target list for twenty-six atomic bombs. Truman relieved MacArthur on April 11, 1951, at least in part because the president believed the use of atomic weapons was so imminent he could not have a loose cannon like MacArthur in charge. MacArthur's successor, Matthew Ridgway, also believed that nuclear weapons needed to be used at once in Korea, and presented to the national leadership a list with even more targets than MacArthur's: thirty-eight. These included, in addition to North Korean cities and military targets, downtown Shanghai.

Preparations to use nuclear weapons advanced so far that in September and October 1951, Air Force bombers actually flew from Okinawa to Korea and dropped dummy nuclear weapons, to verify the feasibility of using them on high-value targets, a series of top secret missions code-named "Operation Hudson Harbor." In the end, it was not any moral qualm that kept Truman from authorizing the use of nuclear weapons in Korea. Instead, Hudson Harbor determined that it was nearly impossible to locate large enough concentrations of enemy troops to make the use of A-bombs worthwhile. The effort proved true the Navy's old argument during the Revolt of the Admirals, that the nation could not rely solely on the power of strategic nuclear bombing for its defense.

As the American Army suffered in Korea and MacArthur faded away, Rickover was choosing a shipyard for his nuclear submarine—the venerable Electric Boat of Groton, Connecticut, a shipyard founded in part by John Holland, the Irish rebel and patriarch of the American submarine fleet. The Pentagon issued a brief one-sentence statement announcing the contract, and a director from the AEC told reporters that the project had crossed the threshold from a fantasy that required a nearly constant lobbying effort to a real life, top secret military endeavor. "From now on," said the director, "you can gauge our progress by the increase in the vagueness of our reports."

Work rapidly accelerated, and Rickover demanded the full dedication of those around him, including the original small group that had journeyed with him to Oak Ridge in 1946. Ruth Masters Rickover wrote that atomic reactors had "elbowed their way into the family and become its most important members," a sentiment undoubtedly echoed by all the wives of all the men that Rickover drove so hard. His battles with the Navy, suppliers, and the federal government were epic, but Rickover slogged on, determined to show it could be done. For the first time, his name became known to the public, mentioned in *Time* magazine on February 26, 1951. The article first quoted AEC commissioner Sumner Pike, who said, "In an attempt to get useful power from atomic fission, we are engaged in the design and construction of a power plant for naval submarines. The design of two practical, though expensive, devices for submarine propulsion is practically complete, and one of them is partly built. It shouldn't be too many years before one or both will be operating in a true submarine." The article went on to say: "There were no further details. But earlier in the week, Navy Captain H. G. Rickover reported on the same project at a highly secret meeting of the congressional Joint Committee on Atomic Energy. Senator Brien McMahon, committee chairman, said afterwards that he was 'both educated and pleased.'"

It was the start of a string of adoring articles the *Time* machine would produce about Rickover. A prominent *Time* journalist, Clay Blair, would even publish what amounted to an authorized, fawning biography of Rickover in 1954. Some conspiracy-minded Navy men were suspicious of *Time*'s motives—the magazine had been one of the Air Force's staunchest allies during the Revolt of the Admirals, and one of the Navy's most strident critics. Never mind that Clay Blair had been a submariner during the war—if *Time* supported Rickover and his submarine, they reasoned, it must be because it would detract from the Navy's carriers. It is far more likely that Clay Blair and his colleagues simply found that Rickover, the curmudgeon who thumbed his nose at the Navy's brass, made great copy. Like MacArthur, he was an American archetype, the brilliant maverick who got things done in spite of his bosses. Americans couldn't help but cheer him on, and for decades reporters like Blair would love Rickover for it.

Despite the favorable publicity in *Time* and the consensus that nearly the entire nuclear submarine effort hinged on his personal resolve, Rickover came up against, not for the last time, the consequences of his maverick ways and his open disdain of Navy tradition. On July 2, 1951, he was passed over for promotion to admiral.

The Navy's system for promoting officers had always been arcane and secretive, and the higher the rank under consideration, the more arcane and secretive the system became. In the months after Rickover was passed over, defenders of the system would trumpet legalistic rationales for the selection board's decision, such as the overall number of EDO admirals, or the more senior men from the Class of 1921 under consideration. However, there was no mistaking the air of retaliation about the board's decision. After Rickover's constant scorn for Navy protocol and custom during twenty-nine cantankerous years on active duty, officers from his long list of enemies had finally formed a quorum and assembled against him in the secret, smoke-filled room of the selection board. And it was much more than a symbolic blow. According to Navy regulations, the selection board met once a year. Any officer passed over for promotion two years in a row was out of a job. Rickover pressed on, without dulling his words to superiors or donning his uniform any more frequently. Rickover's admirers were hopeful that the admirals, having made their point in the 1951 selection board, would promote Rickover in 1952, so the nation could continue on with the vital business of building a nuclear submarine.

On December 12, 1951, the Department of the Navy announced the name of the world's first nuclear submarine. It would pay indirect homage to Jules Verne, who eighty-two years earlier had described a true submarine with an uncanny degree of accuracy. Captain Rickover's boat would carry the same name as Captain Nemo's: *Nautilus*.

On June 14, 1952, the keel of the *Nautilus* was laid at Electric Boat in Groton. Slightly more than three years after the soggy, ill-fated keel laying of the *United States*, the Navy found itself starring in a national celebration, one in which everybody suddenly wanted to talk again about the crucial role of the U.S. Navy. The president of the United States himself, the man who had appointed (and later fired) Louis Johnson, presided over the ceremonial beginning of the ship's construction. Truman began by apologizing to the crowd for mistakenly saying in a speech a week before that the historic submarine would be built in the rival city across the Thames River: New London: "I sometimes get pretty tired of Kansas City taking all the credit for things that happen in Independence, Missouri," he said. "I can understand why the people of Groton should be proud of what is happening here today." Truman then, more eloquently than anyone else on the podium that day, captured the almost magical nature of what Rickover was trying to accomplish.

> *We are assembled here to lay the keel of a Navy submarine, the USS Nautilus. This ship will be something new in the world. She will be atomic-powered. Her engines will not burn oil or coal. The heat in her boilers will be created by the same force that heats the sun—the energy released by atomic fission, the breaking apart of the basic matter of the universe. Think what this means.*

Truman didn't mention Rickover by name, perhaps sensitive to protocol with so many more senior naval officers on the grandstand. The next speaker, Gordon Dean, the chairman of the AEC, showed no such compunction:

> *There are many people who have played a role in the events which have led to this ceremony, but if one were to be singled out for special notice, such an honor should go to Captain H. G. Rickover, whose talents we share with the Bureau of Ships and whose energy, drive, and technical competence have played such a large part in making this possible.*

A giant crane lowered a curved section of steel into the dry dock, where the president climbed down to meet it. Truman pronounced the keel "well and truly laid," and then inscribed his initials on the plate with a piece of chalk. A shipyard worker followed behind him and traced the president's initials for all eternity with an arc welder. (In a strange example of the kind of military sentimentality Rickover disdained, the Navy would use the exact same electrode holder at the keel laying of every nuclear submarine for the next twenty-four years.) Six years after Rickover reported to Oak Ridge without the slightest knowledge of nuclear engineering, construction of the *Nautilus* was under way. Not the least bit circumspect in his most public appearance since being passed over for promotion the previous June, Rickover attended the keel laying ceremony in civilian clothes.

After the ceremony, President Truman saw Rickover in an official car and hurried over to congratulate the diminutive captain in person. Rickover shook the president's hand without standing up or getting out of the car. The unpretentious Truman did not appear offended in the least, but a train of jowly admirals behind him stared on in horror.

Even as the construction of the *Nautilus* in Groton began, a crucial parallel project had begun in Idaho, at the nation's newly established reactor testing station—the construction of a prototype nuclear reactor. Contrary to normal practice, the prototype reactor would be an exact duplicate of the one built for the *Nautilus*, constructed inside a submarine-sized tube. Normally, a prototype of that complexity would be built with its components spread out "breadboard style," for ease of instrumentation, examination, and maintenance. Rickover's aggressive timetable didn't allow for that kind of leisurely analysis. Every lesson learned in Idaho, he mandated, had to directly benefit the submarine in Connecticut. Rickover went so far as to surround the prototype with a 385,000-gallon tank of water, to see what effect the water would have on the radiation and shielding of the operating reactor. If all went according to plan, the reactor in Idaho would be operational slightly before the one in Connecticut, allowing them a few precious months to test the equipment and train the crew before putting them to sea with the machinery on which their lives would depend.

Less than a month after the keel laying, Rickover allowed himself to be pulled away from the frantic work schedule of the *Nautilus* to be awarded a second Legion of Merit, the highest award a naval officer can win in peacetime. The medal was given to Rickover personally by Dan Kimball, who was secretary of the Navy and one of Rickover's strongest supporters. In his commendation, Kimball pronounced that Rickover had "held tenaciously to a single important goal through discouraging frustration and opposition and has consistently advanced the submarine thermal reactor beyond all expectations." Kimball went on to say that the *Nautilus* was the "most important piece of development work in the history of the Navy." He then pinned the medal on the lapel of Rickover's gray civilian suit.

The next day, a panel of nine admirals met in secret and declined to promote the captain a second time. Hyman Rickover had been fired.

JANUARY 3, 1961—9:01 PM

The first people to be aware of a problem at SL-1, other than the three victims, were the firefighters of the National Reactor Testing Station (NRTS). The alarm sounded at 9:01 PM inside fire station #1, the bells ringing a code: long, long, short. An alarm operator read the ADT paper tape recorder, and pulled the file for the appropriate location: SL-1, the tiny Army reactor about eight miles distant.

Egon Lamprecht groaned as the alarm was announced, even as he hustled toward his boots and coat. Technically, the alarm could have been caused by heat, high radiation, a pressure surge, or even a flying projectile. But everyone in the fire station suspected something less dramatic. They had already answered two false alarms at SL-1 that day, each coming from the support building's small furnace room. Egon was twenty-five years old and had been on the force just two years, the youngest man at the firehouse. Consequently, Egon made each trip on the back of their fire engine, exposed to the elements and the freezing temperatures during the nine-minute drive. Egon loved his job, and he considered himself lucky to have it. But even so, he didn't want to catch frostbite just to reset another false alarm.

The crew bundled up and trudged to their vehicles. Of the seven-man crew, one man, the alarm operator, stayed at the station. Three men, including Egon, sat in the open air in the back of the engine with only their firefighting garb for warmth. Two of the more senior firefighters sat in the heated cab up front, and the assistant chief, Walter Moshberger, rode separately in the station wagon—heated, of course. As they sped away from the cozy fire station into the frigid, brittle night, Egon hunched his shoulders and turned his face away from the bitter wind.

Born in 1935, Egon was raised in a small devout community of Mormons in nearby Blackfoot, Idaho. His father had been born in Germany, where he was converted as a young man by some energetic Mormon missionaries. They convinced him to escape the endemic persecution of Mormons in Europe and to join the rest of the faithful in the western United States, as part of "the gathering of the saints" prophesied by Joseph Smith. Lamprecht's father immigrated to the United States about the same time Hyman Rickover's father did in 1900, both men seeking their own kind of religious freedom.

After graduating from high school, and discovering to his disappointment that he couldn't make a living building hot rods, Egon jumped at the chance to work for the NRTS fire department in 1958. He got paid $316 a month, good pay for rural Idaho. As for the reactors he watched over . . . Egon didn't give them much thought. He knew there was something vaguely dangerous inside the buildings that dotted the vast preserve, but everyone seemed so comfortable with them. In fact they all looked just like small, nondescript factories, so it was hard to be too worried. To the young firefighter who had pulled many a victim from a mangled automobile, a far more dangerous aspect of life at the NRTS seemed to be icy weather and the temptation to drive fast on the straight desert highways.

The fire engine and the station wagon sped through the night toward SL-1, the route now familiar. At 9:10, the crew stomped up to the unguarded gate.

During late shifts, it was normal that the gate be unmanned. Procedure had the firefighters walk into a small guardhouse where they could telephone the control room, which was always supposed to be manned, and have one of the SL-1 crewmen come let them through the locked door. The guardhouse also contained a portal radiation monitor, an instrumented doorway through which walked anyone entering or leaving the site. Inside the empty guardhouse, Egon saw for the first time something different from the previous trips they had made to investigate the false alarms in the furnace room. The alarm on the portal's radiation detector was blaring.

And no one answered the phone.

Moshberger, still unfazed, called a security guard at another location, who drove to the facility to unlock the gate with his master key and let the firefighters inside. With the vehicle gate open, the fire engine was able to pull directly up the furnace room, whose alarm they all still assumed was the culprit. They were surprised to see that the furnace room alarm, for the first time that day, was not blinking. The problem at SL-1 was elsewhere.

Moshberger and Lamprecht went into SL-1's small main building, which was quiet and appeared normal. All lights were on and the previously suspect furnace had heated the space to a cozy warmth. Lamprecht was not trained in any way on the operation of a nuclear reactor. As they walked by the unattended control panel, however, he could read and understand the blinking red indicator on the panel: HIGH RADIATION. *Finally suspecting a real problem, Lamprecht and Moshberger retreated from the building.*

Outside, still more befuddled than afraid, they donned Scott air masks and oxygen tanks. As a precaution, they grabbed a handheld radiation detector from the fire truck before walking back into the building. When they reentered, because of the extreme temperature difference between the outdoors and the control room their face masks immediately fogged up. Even with the reduced visibility, Lamprecht noticed that there were still lunchboxes on the table in the lunchroom and coats hanging on the hooks, a sight that unnerved him.

Moshberger, as the assistant chief, had a battery-powered amplified mouthpiece on his air mask that allowed him to shout through the building as they searched: "Anybody here? Anybody here?" No one answered. The radiation detector in Lamprecht's hand clicked steadily and read 25 roentgens per hour.

They moved toward the reactor, which was housed in the adjacent silo-shaped metal building, thirty-eight feet, seven inches in diameter and forty-eight feet high. An unheated, covered stairway curled up the side of the cylinder, connecting the control room to the reactor room, roughly halfway up the silo. The floor at that level was even with the top of the reactor, and the space contained much of the reactor's vital equipment: the tops of the control rod drives, the motor control panel, and the feedwater pumps. Lamprecht and Moshberger climbed the stairs, still hoping to find the three men who were supposed to be on-site, perhaps absorbed in some particularly difficult maintenance or stealing a quick nap on the nightshift.

About halfway up the stairs, the radiation detector pegged at 200 roentgens per hour, the maximum reading for the device. Lamprecht and Moshberger stared at the dial in disbelief. In their cursory training on the detector, they used shielded samples in tiny boxes to make the counter click and the needle deflect slightly. Now the clicks had become so rapid and loud that as they merged into a steady whirr, they reminded Lamprecht, the hot rod enthusiast, of a revving car engine. They retreated down the stairs to get another detector, certain the one in their hands must be broken.

With the new detector, Lamprecht and Moshberger climbed the stairs once again. The new detector also pegged halfway up. Certain now that something was dangerously wrong at SL-1, they hurriedly completed the climb, and both men looked for the first time through the door into the reactor room.

Visibility was surprisingly good. A large ventilation fan in the silo had done a good job of evacuating steam, and all the lights were still on. Neither man had any trouble seeing the first corpse sprawled a few feet away, bloody, twisted, and soaking wet.

Moshberger, who was in front, saw another body, but before Lamprecht could take it all in, the assistant chief had spun around and with a shout ordered him down the stairs. Neither man saw the third body, despite the good visibility. A string of searchers, limited to thirty-five seconds by the extreme radiation, wouldn't see it for two hours. No one thought to look up.

• • •

At the time, Lamprecht's knowledge of radiation mirrored the nation's: he saw it as a relatively benign, mysterious force that needed to be treated with respect, but not fear. His training on the subject was minimal. The firefighters did carry radiation detectors and were trained how to use them. They also wore film badges clipped to their shirts to monitor their personal radiation. But beyond that, Lamprecht didn't know much about what he'd seen, breathed, and otherwise been exposed to the night of January 3, 1961.

Later, as his knowledge of nuclear matters grew during his two decades in the fire department, Lamprecht wondered more about the radiation he had absorbed that night and the possible damage it may have done. He remembered that when his film badge was read after the incident, the "health physicist" told him that he had absorbed 18 R—without bothering to elaborate to the young firefighter what this might mean. And that was a very rough measurement, as the film badges, like the radiation detector in the fire truck, had been designed by people who never envisioned a radiation field as intense as the one at SL-1. Even so, 18 R represented a massive overshoot of the modern limit of 5 rem—and that's the limit for a full year of exposure, not the roughly thirty minutes Lamprecht had spent inside the radiation field. It also bothered Lamprecht that in all the years after the accident, no doctors ever examined him for long-term effects from the blast, or even discussed with him what those effects might be.

Once, decades after the accident, Lamprecht went to pull his health record at the Central Facilities Area to review his lifetime radiation exposure. He wondered if there might be insightful notations from a doctor in the margins, or some medical opinion about the possible lasting effects of the most memorable night of his career. After a brief search, a confused clerk returned to the front counter and told Lamprecht that his exposure record had disappeared. So had Moshberger's.

chapter 3

IDAHO

Hyman Rickover wasn't the first to find a use for the wide open spaces of southeastern Idaho. For much of its history, sheer desolation has been the region's most coveted natural resource. The lava flows that were the defining characteristic of the geography were the result of volcanic eruptions that ended as recently as 2,100 years ago, when ancestors of the native Shoshone population actually witnessed the eruptions. A Shoshone legend tells of a giant snake who squeezed a mountain until it exploded. If that wasn't a cultural memory of volcanism, perhaps it was a prophecy of the work that would be done in their part of Idaho in middle of the twentieth century.

Until 1942, the area appealed only to a small but steady stream of Mormon settlers, the nation's first devotees of western isolation. They built small, industrious communities such as Blackfoot, Arco, Pocatello, and Idaho Falls, free from discrimination and free to practice the more controversial tenets of their faith. Their peaceful seclusion ended with the attack on Pearl Harbor, an event that made American desolation suddenly desirable to a variety of constituencies.

First came the Minidoka internment camp, hidden away on thirty-three thousand acres of federal land twenty miles east of Twin Falls, Idaho. The camp became home to nine thousand relocated Japanese Americans who were allowed to bring with them only what they could carry from their homes in Washington, Oregon, and Alaska. The Morrison-Knudsen Company of Boise was awarded the $3.5 million contract to build the camp's drab, low-slung barracks and command buildings. The Idaho location was chosen to put the evacuees out of range of both invading Japanese hordes on the West Coast and any potential sympathizers in the American public at large, two groups regarded as equally dangerous by the War Department. Arthur Kleinkopf, the superintendent of education at Minidoka, wrote in his diary a description of the kind of lonely feelings the Idaho landscape could evoke.

> *These people are living in the midst of a desert where they see nothing but tar paper covered barracks, sagebrush, and rocks. No flowers, no trees, no shrubs, no grass. The impact of emotional disturbances as a result of the evacuation procedures, plus this dull, dreary existence in a desert region surely must give these people a feeling of helplessness, hopelessness, and despair which we on the outside do not and will not ever understand.*

The federal government found other uses for the Idaho desert. A naval ordnance plant was hurriedly constructed in Pocatello, just months after Pearl Harbor. Just as with the internment camp, the government wanted an isolated locale far from any potential West Coast invasion or saboteurs. The plant's mission was to overhaul the massive, worn guns from the Navy's biggest ships; the guns were shipped by rail to Idaho, where defense workers relined the barrels, enhancing their accuracy and safety.

To test their work, the ordnance plant needed a firing range of mammoth proportions. The biggest guns they worked on, the sixteen-inch guns from battleships, could hurl a 2,700-pound shell a distance of twenty-three miles. The firing range needed to be flat, and, obviously, unpopulated. The Navy found exactly what it was looking for about fifty miles north of Pocatello, and commandeered 271 square miles of desert to build the Naval Proving Ground. Like the Minidoka internment center and the Pocatello ordnance plant, every building at the proving ground was constructed by Morrison-Knudsen, whose executives must have been gaining a unique perspective on the course of World War II.

The test firing of guns began on November 20, 1943. The workers in Idaho, many of them women, would test hundreds of guns, of all calibers, before the war was over. They were constantly reminded that their work was crucial to the war effort, that the great American fleet needed the gun barrels to make the seas safe again for Democracy. These Women Ordnance Workers, or "WOWs," would have been shocked to learn that soon a new technology, born in another desert and refined in their own corner of Idaho, would make both those big ships and big guns obsolete.

An atomic bomb was detonated near Alamogordo, New Mexico, on July 16, 1945, a test code-named "Trinity" with typical aplomb by Manhattan Project chief Robert Oppenheimer. On August 6, "Little Boy" was dropped on Hiroshima, and three days later, "Fat Man," a twin of the weapon exploded in New Mexico, exploded over Nagasaki. Even before the Japanese surrender on August 15, there was speculation about how the power of the atom bomb might be peacefully channeled. An August 9 editorial in the *New York Times* on the bomb speculated that fission might become a new form of aircraft power, and quoted a scientist who predicted an atomic automobile engine, "no larger than a brick."

Fat Man used plutonium (Pu-239) as fuel, and a complex set of explosive lenses to focus that fuel into a perfect sphere, at which density it reached critical mass and exploded. Little Boy used U-235 as its fuel and was of the much simpler "gun" design: it simply shot one slug of uranium into another to reach critical mass. The different concepts led to the distinctive profiles of the two weapons. Trinity was used to test the firing of the implosion device because the implosion-style weapon was much more complex, while the simple gun style was seen as almost foolproof. Additionally, Little Boy had nearly used up the country's entire supply of U-235—there wasn't enough left for a test.

Plutonium does not exist in appreciable quantities in nature. It is created in a nuclear reactor. During World War II, the production of plutonium was the business of the massive reactors along the Columbia River, in Hanford, Washington. U-235, on the other hand, must be painstakingly separated from naturally occurring U-238, work that took place at the Manhattan Project's other monumental facility, in Oak Ridge, Tennessee. During the course of the Manhattan Project, it became clear that plutonium was better bomb-making material. Not only was it easier to manufacture, it also took less plutonium to achieve critical mass, and its lower rate of spontaneous fission made it easier to put the bomb together without "fizzling," a premature, incomplete reaction. A nuclear reactor was required to make plutonium. Inside a nuclear reactor fueled by uranium, the manufacture of plutonium is almost inevitable. This is the crucial link between nuclear power and nuclear bombs, a link that has influenced the development of nuclear energy from its earliest days.

On August 1, 1946, President Truman signed the Atomic Energy Act into law. The law took control of the nation's atomic reactors away from the military, and gave it to a new civilian organization known as the Atomic Energy Commission. The AEC, or at least the regulatory arm of it, would be renamed the Nuclear Regulatory Commission in 1974. In 1946, however, even calling it the Atomic Energy Commission was euphemistic. Creating usable energy from fission was a fantasy at the time, and in light of the group's actual priorities, it would have been far more accurate to call the group the Atomic Bomb Commission.

The AEC was ruled by a powerful group of legislators known as the Joint Committee on Atomic Energy. One of the first jobs for both groups was to choose a location for a National Reactor Testing Station, an area, far from any population center, where scientists could test the dangerous outer limits of nuclear science. A number of western towns competed vigorously for the honor. After a lengthy evaluation the commissioners chose the most isolated parcel in the federal government's vast inventory of land holdings: the Naval Proving Ground of southeastern Idaho. On February 18, 1949, the Atomic Energy Commission announced its decision, much to the dismay of the Navy, which didn't want to lose its proving ground and fought the decision until the end. Also bitterly disappointed were the town fathers of Fort Peck, Montana, who had hoped their town would be the focus of the nation's nuclear experimentation.

Rickover would not be the only tenant of the NRTS, although he was one of the first, breaking ground on the *Nautilus*'s prototype reactor in August 1950, almost two years before Truman presided at the submarine's keel laying in Connecticut. Soon enough, all three military branches would be represented in Idaho. A fourth presence was the civilian sector, represented in large part by Argonne National Laboratory.

Argonne was a direct descendant of Enrico Fermi's pioneering work at the University of Chicago for the Manhattan Project. Fermi's Chicago lab had been discreetly named the "Metallurgical Laboratory" or "Met Lab" during the war. In 1946, the lab was moved from the city to a more distant, presumably safer site near the Chicago suburb of Palos Park. The lab took its new name from the nearby Argonne Forest section of the Cook County Forest Preserve, which in turn took its name from the World War I battle in France.

Argonne could take credit for the world's first controlled nuclear reaction, which took place in Fermi's CP (Chicago Pile)-1 reactor, inside a racquets court beneath the west stands of the University of Chicago's football stadium on December 2, 1942. Since the war's end, Argonne had been headed by Walter Zinn, one of Fermi's closest colleagues. The complete dedication of almost all enriched uranium to weapons production prompted Zinn to make a seemingly outlandish proposal to the AEC—he said he could build a reactor that would actually create more fuel than it consumed, generating usable power while at the same time "breeding" plutonium, both for the bomb makers and for its own fuel. Zinn's proposal got the government's attention, and it became one of the first reactors, along with Rickover's submarine reactor, constructed in Idaho. As the only redbrick building on the Idaho site, Zinn's facility seemed almost to pay architectural tribute to Argonne's birthplace beneath a college football stadium. Inside that building, Zinn would accomplish one of those rare early milestones of nuclear power that had nothing to do with Rickover.

The principle behind Zinn's breeder reactor was simple. A small core of enriched uranium (U-235) would be surrounded by a shell of the relatively plentiful, unenriched natural uranium (U-238), the kind of uranium that can be dug out of the ground but is very reluctant to fission. As the internal core became critical, a portion of the neutrons it produced would proceed into this outer shell of natural uranium, which would, through a series of nuclear reactions, transform some of the U-238 into plutonium. In this manner, Zinn's machine would actually create more fuel than it used. He called it EBR-1, for "Experimental Breeder Reactor 1." Smart-aleck contractors in Idaho would erect a sign along Highway 20 that read WARNING: DO NOT DISTURB BREEDING REACTORS.

The breeding of nuclear fuel was the exciting part of the experiment to scientists like Zinn. The generation of power by the reactor was almost an afterthought. It had long been known that nuclear reactions created heat, and man had for eons known a number of ways to transform heat into other useful forms of energy. Nonetheless, no one had ever managed to do it with a nuclear reactor before Zinn.

On August 24, 1951, EBR-1 went critical for the first time. On December 20, after several months of testing, Zinn and his team from Argonne increased the reactor's power and began to heat a liquid compound of sodium-potassium that flowed through and around the fuel. Sodium-potassium, represented by the chemical formula NaK and referred to as "nack," ran through a steam generator, where its heat was transferred to water, which soon boiled. The resulting steam turned a turbine and an attached electrical generator. The generator was hooked up to four very pedestrian lightbulbs hanging from a handrail. As soon as the switch was thrown, the four bulbs glowed brightly, the first electrical devices ever powered by nuclear energy.

Zinn, whose real priority was the fuel breeding taking place unseen inside the core, didn't make any lofty pronouncements at the time. He recorded the event simply in the log: *Electricity flows from atomic energy. Rough estimate indicates 45 kw.* Perhaps after reflecting more on the milestone, the next day Zinn wrote in chalk on the concrete wall of EBR-1 all the names of those present for the event, along with a doodle of a devilish head blowing a cloud of steam. That same day, EBR-1 began supplying all the electricity for the facility.

Zinn and Rickover did not get along. Their personal battle was reflected in an enduring wall of tension between Argonne and the Naval Reactors Branch of the AEC. The two organizations worked together at times, by necessity, in an era when nuclear resources and expertise were severely limited and both organizations were ruled, to a large extent, by the same federal bureaucracy. Nonetheless, they both made a practice of downplaying the other's contributions, a practice that extends through the decades to the official and unofficial histories written for both organizations. To Rickover, Zinn and his colleagues epitomized the academic attitude toward nuclear power: fusty, plodding, and not all that concerned with practical results. Zinn saw Rickover as many others did: a bullying, controlling autocrat who harangued and exiled all those who dared challenge him. In later years, some in the nuclear power community would reduce the battle between Zinn and Rickover to a battle between the breeder reactor and the types of reactors Rickover wanted to build for Navy ships. Rickover did, in fact, see the breeder as a needless distraction from his quest. But the real conflict between the two men was more fundamental. Each was a strong, smart leader who thought he was best qualified to direct the future of American nuclear power.

(Within a few years, events would seem to validate whatever negative feelings Rickover had about the breeder reactor. EBR-1 would melt down on November 29, 1955, the first unintentional meltdown of a nuclear reactor. The accident occurred when rods weren't inserted fast enough during testing. Fermi 1, in Monroe County, Michigan, a breeder reactor directly inspired by Zinn's work and the dream of unlimited fuel, melted down on October 5, 1966. That incident would inspire the 1975 antinuke classic, *We Almost Lost Detroit*, by John G. Fuller. Advances in enriching uranium also made the need for breeding reactors less pressing.)

Rickover broke ground on the *Nautilus* prototype plant in Idaho in August 1950—a full year before the submarine had even been officially ordered from Electric Boat. He would masterfully play off his dual roles within the Navy and the AEC, strong-arming the AEC into building the reactor, and then telling Congress what a waste it would be for the Navy not to build the submarine now that the AEC had constructed a perfectly good power plant.

By the time construction began, the water-cooled Westinghouse design had become the clear winner over the sodium-cooled General Electric plant. NaK, the same coolant used in Zinn's EBR-1, had initially seemed very promising, with many of the traits of an ideal reactor coolant. NaK was liquid at room temperature. It did not absorb any of the precious neutrons flying through the core and had superior heat transfer properties to water. In almost every other respect, the engineers found, water was a better reactor coolant. NaK actually burned when exposed to air. Worse, it exploded when it came into contact with water—not a good characteristic for anything flowing through the veins of a submarine. While the Navy would persist in trying to use NaK for a number of years, it was eventually deemed impractical for naval plants. Rickover would comment in later years that if the ocean were made out of liquid sodium, "some damn fool would make a submarine reactor cooled with water."

So Westinghouse, under Rickover's watchful eye, constructed the prototype in the Idaho desert, an exact replica of the plant being built for the *Nautilus* in Connecticut. Rickover's goal was to have the prototype be just one year ahead of its twin at Electric Boat, so that the lessons of the prototype could be immediately applied to the construction of the ship. The reactor would have many names, but one of the first was "STR," for "Submarine Thermal Reactor." The word *thermal* describes neutrons that are slowed down, an energy level at which they are more likely to cause fission in the U-235 fuel. (Zinn's breeder reactor, by contrast, used the fast neutrons more amenable to converting U-238 to plutonium.) Soon Rickover's plant came to be known as "S1W," the "S" for submarine and the "W" for Westinghouse.

While the specifics of building the reactor were immensely complex, the overriding principles of the S1W plant were relatively simple, a necessity for a plant that would be run at sea by Navy men, not in a laboratory staffed by chemists and physicists. The fuel was enriched uranium. The control rods that ran vertically through the core were made out of hafnium, an element very prone to absorb neutrons. With the control rods fully inserted into the core, the hafnium essentially poisoned the reaction: the vast majority of the neutrons were absorbed by the hafnium, not leaving enough to collide with the uranium nuclei and create a chain reaction. The control rods were raised and lowered by small electric motors and gears: the control rod drives. As the rods rose, they compressed springs. If any part of the control rod drive failed, gravity and the springs would thrust the rods to the bottom of the core, immediately shutting down the core in an event called a "scram" since the earliest days of nuclear power. One origin myth around the term held that at Fermi's pile at the University of Chicago, a man with an ax was stationed by the rope that pulled the control rod. In the event that the reactor ran amok, he was to cut the rope, which would let the rod fall back into the core and (hopefully) shut it down. That man, according to the legend, was called the "Supercritical Reactor Ax Man," abbreviated "SCRAM."

To start up S1W, the rods had to be slowly lifted, removing them from the core. Fewer neutrons would be absorbed by the hafnium of the rods, and more neutrons would collide with the uranium fuel. When a neutron collided with a uranium nucleus, it would fall apart, or fission, creating a variety of new elements depending on exactly how the uranium atom split. While falling apart, the uranium nucleus would also release one or more neutrons, which would then collide with other uranium nuclei, sustaining the chain reaction. When exactly enough neutrons were being liberated to sustain the chain reaction, the reactor was said to be critical, the desired condition. Too few neutrons and the reactor was subcritical; too many and it was supercritical. To reduce reactor power slowly, rods could be shimmed inward, reducing the number of neutrons available for the chain reaction.

In addition to releasing neutrons, the fissioning atoms also released the considerable energy that had been holding the uranium nuclei together. Some of this energy was released in the form of heat, which was absorbed by the water flowing through the core. In this capacity, water was the reactor's "coolant." In the way that collisions between water molecules and neutrons slowed the neutrons down to thermal speed, water was also the reactor's "moderator." Some early reactors, like Fermi's CP-1, were moderated by the carbon atoms in graphite bricks rather than water.

The water flowing through the reactor and absorbing heat, the "primary" side of the plant, was kept at high pressure so that the water would stay in its liquid form, despite being heated to hundreds of degrees above its normal boiling point. This system gave Rickover's reactor another one of its many names, the pressurized water reactor, or PWR. The hot, pressurized water from the reactor was pumped through the pipes of a steam generator. Water, at lower pressure, flowing over these pipes would absorb the heat, turning to steam. This was the "secondary loop," which was nonradioactive. This design had two principal advantages over a reactor that boiled water and turned turbines in a single loop. For one thing, since the primary side never boiled, the hot reactor core was always covered with water, an important safety consideration. Secondly, this design spared the operators and engineers the problems of radioactive steam and radioactive turbines.

A traditional Navy engine room burned some kind of fossil fuel. The heat from this combustion was used to boil water. The steam was used to spin a turbine that could either generate electricity, or, through a series of reduction gears, turn the screw that propelled the ship forward. The steam was then cooled, returned to its liquid state, and pumped back through the boiler where the cycle began anew.

In a nuclear engine room, the heat source of burning fuel was replaced by the nuclear reactor. This had the enormous advantage for submarines of not requiring oxygen. Steam from steam generators would flow to turbines, and from there on, the nuclear submarine's engine room looked very much like a traditional Navy engine room. Of course, there were huge new challenges to overcome, however simple the thermodynamics seemed to be. Enormous amounts of shielding had to be put in place to protect the crew from radiation. Materials, such as the hafnium of the control rods and the zirconium that held the fuel elements together, had never before been mined, purified, or machined on an industrial scale. Everything had to be tested to ensure it could withstand the heat, radiation, the pressure, and even the motion of a submarine plant. All of this was invented on the fly, in an incredibly short period of time, as the submarine engine room took shape in the Idaho desert, a thousand miles from any ocean.

As if building a nuclear ship weren't challenging enough, Rickover also had the job of building the crew who would operate her. The early officers he recruited made up an eclectic and impressive group. Eugene "Dennis" Wilkinson, the man who would be the first commanding officer of the *Nautilus,* was a demonstration of Rickover's preference for non–Naval Academy men; he viewed academy graduates as too inflexible and lacking in creativity. (Railing against the teaching methods of the academy would be a hobby of Rickover's for decades, perhaps exacting revenge for the four friendless years he'd spent as a midshipman.) Wilkinson had graduated from San Diego State University, and had taught college chemistry as a nineteen-year-old. Alex Anckonie, who would also go on to command the *Nautilus,* was an observant Muslim and once had the unenviable task of asking Rickover for five weeks off to make his pilgrimage to Mecca. Rickover granted the request.

Rickover personally interviewed every single officer in the naval nuclear program. By his own calculations near the end of his career, he had interviewed more than fourteen thousand men, and for every one of them it was an unforgettable moment in their lives, a rite of passage for more than a generation of nuclear-trained officers. Rickover almost always accosted the applicant in some way, attacking his grades, his choice of classes, or his overall decision-making ability and good sense. Rickover sawed down the front two legs of a chair in his office to make the midshipmen and junior officers even more uncomfortable during the interview as they slid forward in their seat. Men with the audacity to plan on marriage and family were denounced by

Rickover as "damn nest builders." Some were led away to "the box," a windowless office where they were abandoned after being asked to contemplate what they had done to make the admiral so angry. The classic Rickover interview story begins with scalding insults from Rickover, and ends with admission into the nuclear program. One applicant told the story of Rickover saying that he liked his necktie and actually asking if he could have it. When the applicant refused after repeated requests, Rickover smiled and showed him a cabinet filled with the ties of less courageous nuclear hopefuls. The interviews did more than give Rickover the final say on every officer in his program. The process also made it clear to every nuclear-trained officer that he was entering a unique realm within the Navy, a realm that belonged wholly to Rickover.

Many applicants failed to see the charm in Rickover's methods. To them he could seem arrogant, petty, and even sadistic. Young men were told to call their fiancées on Rickover's phone and cancel their weddings. Naval Academy men, in particular, were subjected to rants about the limitations of their education. Sometimes men even declined to enter the nuclear Navy specifically because of what they'd seen of Rickover in the interviews. One of these was Elmo Zumwalt, who sat before Rickover in 1959, not as a cowering midshipman but as an experienced commander, in line for command of a nuclear-powered surface ship—if he could pass muster with Rickover. Over the course of a lengthy interview, Rickover called Zumwalt "a stupid jerk," "greasy," and "one of those wise goddamn aides." As was often the case, the abuse was not indicative of a failing interview; Rickover actually selected him for the program. Zumwalt, however, had decided he didn't want to play on a team coached by Rickover, and continued his career on conventionally powered surface ships. He became the youngest chief of naval operations in the history of the Navy, at the age of forty-nine, in 1970.

One of Rickover's earliest recruits was a promising young engineer from the Naval Academy class of 1945 named James Carter. The future president of the United States would go to work for Rickover in October 1952, only to resign his commission a year later when his father died and Carter felt compelled to leave the Navy and return to Georgia to run the family farm. Carter would describe Rickover as having a profound effect on his life, "perhaps more than anyone except my own parents." He also said candidly, in a sentiment that would sound familiar to all the thousands of men who would work for Rickover, "I do not remember in that period his ever saying a complimentary word to me."

Carter's single year in the nuclear Navy was long enough to place him at the scene of one of the world's first nuclear accidents. A powerful Canadian research reactor called NRX, located near the Chalk River, in Ontario, had been built during World War II ostensibly for research, but also as an integral part of the fledgling British nuclear weapons program. On December 12, 1952, a series of operator errors caused a set of control rods to be raised inadvertently, raising power so high that part of the plant was destroyed and millions of gallons of radioactive cooling water were released into the building. The United States sent a small group of nuclear-trained specialists to assist in disassembling the core, a group that included Lieutenant Jimmy Carter. In ninety seconds on the scene, Carter absorbed a year's maximum permissible dose of radiation. So great was the radiation that for a year afterward, Carter's feces and urine were analyzed for radioactivity.

It was on July 8, 1952, that the Navy chose not to promote Rickover for the second time, seemingly ending his naval career. At that time, while the keel had been laid for the *Nautilus,* the prototype reactor was not yet complete. Rickover was on a mission to build the *Nautilus,* and had become adept at overcoming daunting hurdles. Now the Navy itself had become one of those hurdles. If Rickover had few friends in the higher reaches of the Navy, he did have friends in Congress, and in the press. While still hammering the submarine reactor into existence, Rickover declared war on the Navy, and like any great military leader, he pressed the advantages he had.

Some of his most powerful friends in the media were the men of Time, Inc., especially Clay Blair, the World War II submariner who had adopted the cause of Rickover as his own. On August 4, 1952, just weeks after Rickover's second selection board, *Time* fired the first salvo in a fusillade with an article about Rickover titled "Brazen Prejudice." While the title referred to the prejudice of the old Navy against technically adept specialists, there was no mistaking the implication that anti-Semitism had influenced the decision of the Navy brass. *Time* summarized the consequences of the selection board's decision, and the Navy's refusal to acknowledge the value of technical expertise:

> *The Navy's failure to recognize this in Rickover's case promises to cost it a brilliant officer who developed the most important new weapon since World War II. Rickover has now been passed over twice, and has completed 30 years' service. This means that, barring unlikely special action, he must retire—at age 52.*

At the same time, Blair began working on what was essentially an authorized, fawning biography of Rickover, written in Rickover's office, typed by Rickover's secretary, and thoughtfully edited by Rickover's wife, Ruth. While the book, The Atomic Submarine and Admiral Rickover, wouldn't be published until 1954, Rickover, Blair, and the rest of his team made sure all parties were aware of its imminent publication.

While *Time* was coy in its allusions to possible anti-Semitism in the Rickover controversy, nationally syndicated columnist and radio personality Drew Pearson was less subtle. (James Forrestal, before killing himself, was one of the vitriolic Pearson's favorite targets. In Bethesda Naval Hospital, before his suicide, Forrestal's psychiatrists found that his weekly anxiety attacks actually coincided with Pearson's Sunday night radio broadcasts.) With his trademark indignation on full display, Pearson railed continuously against the Navy's "Brass hats" who had blackballed Rickover "because of his religion." With Pearson and *Time,* Rickover had two of the nation's most powerful media voices firmly in his camp.

The Navy found itself in the middle of a huge public relations mess. Rickover and his allies had successfully portrayed him as the victim of a fusty, jealous, anti-Semitic Old Boys Club of admirals. Had this been the limit of Rickover's offensive, however, the admirals may have endured. They had withstood extremely bad press before, after all, as when war hero Omar Bradley accused them of being "fancy dans" during the Revolt of the Admirals just three years before. There were many in gold braid who would have happily endured a few weeks of feisty newspaper stories if it meant the end of Rickover. Unfortunately for the fancy dans, Rickover had also made a number of highly influential friends in Congress.

Unlike many military men, with their institutional disdain of politicians, Rickover had carefully cultivated relationships with the men on the Joint Committee on Atomic Energy and many other powerful elected officials. In his campaigning for the nuclear submarine, he had first testified before Congress on February 9, 1950, just months after the shocking announcement of the Soviet atomic bomb. Rickover was an impressive speaker, convincing and confident as he told the congressmen what they desperately wanted to hear, that his nuclear submarine would give the United States a decisive weapon against the Soviet menace. It was the beginning of a lasting mutual admiration society of Rickover and Congress, one that would endure for decades. Rickover's friends in Congress included such powerful men as Senator Henry "Scoop" Jackson of Washington, Representative William H. Bates of Massachusetts, and Representative L. Mendel Rivers of South Carolina. (All three would, not coincidentally, eventually have nuclear submarines named in their honor.) In the overheated climate of the Cold War, these men saw correctly in Rickover a man who could get things done. They were not moved by the Navy's appeal to tradition and the sanctity of the selection boards. Equally unmoving were the arguments that Rickover's job as head of Naval Reactors was a "Captain's billet," and that Rickover was replaceable. No one who knew anything about the man believed that. The Korean War was raging and the Soviets had the bomb—now was not the time to kick out military leaders with the vision and the skill to bring decisive weapons into being.

Rickover's allies in Congress fired their own decisive weapon on February 26, 1953. The Senate Armed Services Committee, headed by Rickover ally Senator Leverett Saltonstall of Massachusetts, announced it would delay the promotions of all thirty-nine Navy captains who had been selected for promotion to admiral until an investigation of the entire selection board system was complete. It was the first time since the selection board system had been adopted, in 1916, that the Senate had failed to accept the Navy's list. If you won't promote Rickover, the Senate told the Navy, then we won't promote any of you.

The Navy promptly issued its unconditional surrender. Secretary of the Navy Robert Anderson had just come into town with the recently inaugurated Dwight Eisenhower, and like Eisenhower, he wanted nothing more than to put the Rickover mess behind him. Anderson convened a special selection board to retain Rickover one year more, so that he could be considered (and promoted) by the July 1953 selection board.

On March 30, 1953, the S1W reactor in Idaho went critical, sustaining its first chain reaction—a landmark in the life of any nuclear reactor. Typically, Rickover was not awed by the milestone, as criticality represented more of a scientific goal than an engineering one, a moment detectable on delicate instruments but not by the accomplishment of any real work. S1W was kept at just this power level for weeks while reams of data were gathered.

Finally, on May 31, 1953, the reactor was deemed ready for a higher power level. Rickover invited Representative Thomas Jefferson Murray of Tennessee, the first engineer to serve on the Atomic Energy Commission, to do the honors. Inside the hull of S1W, the perfect simulation of the nuclear submarine, surrounded by 385,000 gallons of water, Congressman Murray slowly turned the throttle counterclockwise, opening the valve that admitted steam to the prototype's main engine for the first time. From there, Murray and Rickover exited the hull and walked down a wooden staircase to the back of the building, where they watched the giant shaft slowly turn. Now, the reactor was doing something more important than creating just enough neutrons to sustain a chain reaction. It was producing real power, the kind of power that would soon propel a submarine through the ocean for weeks, or even months, at a time. In later years, Rickover spoke about the moment with uncharacteristic sentimentality:

> *I haven't experienced real elation many times in my life, but I recall two such times clearly. Once was when I learned that I might go to the Naval Academy and receive a college education, a dream that had previously appeared out of reach. The other time was when I first saw the turning of the propeller shaft of the Nautilus prototype, and I knew we had finally proved for the first time that the atom could do a significant amount of useful work.*

On June 25, Rickover had S1W brought to full power for the first time. After twenty-four hours, the engineers reported to Rickover that they had gathered sufficient data during the test and were ready to shut the reactor down. Rickover, sensing the significance of the moment, vetoed that immediately, and over the strenuous objections of his top officer on-site, Rickover decided on the fly to simulate a submerged transatlantic run. He had a chart of the Atlantic hung outside the prototype hull, and each off-going crew proudly plotted how far they had propelled the imaginary submarine. Inevitably there were problems—generators sparking, instruments failing, and condensers losing vacuum—but in true Navy fashion, the men kept everything running for ninety-six hours, until the line on chart hit landfall at Fastnet Rock, Ireland.

Two days later, a board of admirals met in Washington and grudgingly admitted Rickover to their very exclusive fraternity.

As Rickover became an admiral and finished willing the *Nautilus* into existence, it seemed to many in the public that the promise of nuclear energy might finally be fulfilled. After the war, magazines and the popular press deluged the American public with the utopian promise of unlimited power. Lewis Strauss, chairman of the Atomic Energy Commission, famously forecasted that electricity generated by the atom would be "too cheap to meter" in a speech to the National Association of Science Writers. A *Newsweek* article made even Strauss seem timid:

> *In a relatively short time we will cease to mine coal. The gasoline service station will disappear from the road sides . . . Our automobiles eventually will have atomic-energy units built into them at the factory so we will never have to refuel them . . . Steamships and locomotives operated by atomic energy will be practical in a short time. So will very large airplanes . . .*

King Features published in 1949 the comic book *Learn How Dagwood Splits the Atom!*, with a foreword written by General Leslie Groves, head of "the great organization that developed the atomic bomb." While predictably enthusiastic (Blondie: "My goodness, aren't atoms wonderful!"), perhaps the most surprising thing about the comic book is its nuanced description of the prospects for atomic energy. Mandrake the Magician explains to Dagwood the complexities of harnessing the power of the Bomb:

> *The radiation is bad because it must be guarded against when an atomic pile is used to produce energy. Atomic power for an automobile is, therefore, not very probable, because the shield for absorbing the harmful radiation would weigh many times more than the automobile itself.*

Mandrake's sensible caution would for a decade be absent from any Air Force evaluation on the prospects for a nuclear-powered plane.

Part of the general public's optimism was driven by an almost religious belief that the terrible destructive power of the bomb had to be balanced by some force of good. By the time Rickover was promoted in 1953, this was becoming harder and harder to believe, especially with the detonation of the Soviet bomb in 1949, and the well-publicized detonation of the American H-bomb on November 1, 1952, on the Eniwetok atoll, a bomb 450 times more powerful than the one dropped on Hiroshima. Scientists who initially had shared in the optimism about nuclear science were discouraged by the ever-increasing power of atomic weapons, and that all nuclear materials and research were still controlled by a secretive federal government. Bomb shelters became fallout shelters as a nervous public realized no one could survive a direct atomic hit; schoolchildren were taught in civil defense school movies to be ready for atomic war, and to "duck and cover" by an animated turtle named Bert. It was beginning to appear to many that the world might have been better off had the secrets of the atom never been discovered.

President Eisenhower, eager to preserve America's enthusiasm for nuclear science, announced a major program that would develop uses for the atom beyond bomb making. He declared this in the landmark "Atoms for Peace" speech, delivered to the United Nations General Assembly in New York on December 8, 1953. In front of Secretary-General Dag Hammarskjōld, President of the General Assembly Mme. Vijaya Pandit, and the packed General Assembly Hall, Eisenhower began by outlining the stark realities of the nuclear age, emphasizing the word *all* each time it occurred:

> *Today, the United States' stockpile of atomic weapons, which, of course, increases daily, exceeds by many times the explosive equivalent of the total of all bombs and all shells that came from every plane and every gun in every theater of war in all the years of World War II.*

Over the next twenty-five minutes, Eisenhower outlined a variety of ways that the world might put nuclear science "into the hands of those who will know how to strip its military casing and adapt it to the arts of peace." These included, principally, the creation of the International Atomic Energy Agency under U.N. auspices, which would develop an international bank of uranium and other fissionable materials for use by scientists and engineers developing peaceful purposes for nuclear power. Eisenhower completed his speech with a lofty statement appropriate for the idealism of the United Nations, an institution only eight years old at the time:

> *To make these fateful decisions, the United States pledges before you—and therefore before the world—to devote its en-*

> *tire heart and mind to find the way by which the miraculous inventiveness of man shall not be dedicated to his death, but consecrated to his life.*

Eisenhower gave the Atoms for Peace speech before a single civilian nuclear power plant had even been started, Zinn's four lightbulbs notwithstanding. Finding peaceful uses for nuclear power would be more challenging than anyone envisioned, and Eisenhower would find that in order to strip nuclear power of its "military casing" he would need the help of a Navy admiral. Ground was broken on the world's first commercial nuclear power plant in Shippingport, Pennsylvania, on September 6, 1954, with Duquesne Light & Power in an unequal partnership with Admiral Hyman Rickover. Many in industry and in the AEC were resentful that an admiral would play such a prominent role in what was ostensibly a civilian enterprise, but in the end they all recognized that Rickover, and perhaps only Rickover, could get the job done.

One of the stranger manifestations of the Atoms for Peace ethos was Project Plowshare: the government program promoting the use of nuclear explosives for large-scale excavations. Edward Teller, the program's biggest backer, called it "geographical engineering." A number of demonstration sites were seriously considered, including a new Panama Canal and a harbor for oil tankers in northern Alaska, to be blasted out of the ocean by five simultaneous nuclear explosions. Teller, with characteristic hubris, suggested they carve the harbor in the shape of a polar bear. While the polar bear–shaped harbor never came to be, Project Plowshare did live on with the persistence and massive budgets that would characterize so many nuclear endeavors. Between 1957 and 1974, the Atomic Energy Commission poured $770 million into Plowshare. The program's pinnacle was Project Sedan, a 104-kiloton blast on July 6, 1962, at the time the largest nuclear explosion ever in North America. Designed to show the potential of nuclear explosives for earthmoving, the bomb, detonated 635 feet below ground, formed a perfectly round, 1,200-foot-diameter crater by ejecting twelve million tons of sand into the clear Nevada sky.

A month after the Atoms for Peace speech, on January 21, 1954, the *Nautilus* participated in one of the many ceremonies the Navy demands of its new warships, ceremonies Rickover strategically used to bolster his program. While Truman had spoken at the *Nautilus*'s keel laying, first lady Mamie Eisenhower sponsored the christening.

In 1952, when President Truman had presided over the keel laying, it took an active imagination to look at the curved piece of steel he was initialing and envision the submarine that would grow from it. Now, it was fully formed, although thick fog at first obscured the full length of it from the fifteen thousand shivering spectators in attendance. During the speech by Lewis Strauss, chairman of the AEC, the fog suddenly lifted, and some in the crowd actually gasped at what they saw. Submarines, which like icebergs are normally mostly hidden beneath the waterline, always look startlingly large out of the water, and the *Nautilus* truly was, as Strauss said at that moment from the podium, "something new under the sun." She was 323 feet long, fifty-six feet longer than the *S-48* on which Rickover had served. Her displacement, the truest measure of any ship's size, was 4,092 tons, nearly three times as large as Rickover's old boat. The boat still had the profile of an older submarine, a teardrop-shaped cross-section, although nuclear power would soon make that silhouette obsolete. Because of nuclear power, submarines would forever after be designed entirely around their underwater performance, which was optimized by a perfectly circular hull. The dream of the true submarine had by then thoroughly captured the public imagination. The *Nautilus* unit insignia—a cartoonish fighting submarine standing on its tail, the electrons of an atom swirling around it—had been created by Walt Disney.

Rickover was, of course, on hand. He wore his admiral's uniform for the occasion, shocking some of the men who had worked for him for years and never seen him so dressed. From his seat on the podium he listened impassively as speaker after speaker mentioned him by name. Finally, with all the speeches complete, Mamie Eisenhower walked up the narrow platform that led to the bunting-decked bow of the giant submarine. As she hefted the full bottle of champagne, made heavy by a chrome sheath, a yard worker perched overhead shouted down, "Be sure and hit it hard, Mrs. Eisenhower." She gamely swung the bottle and smashed it against the bow. As the ship slid down the greased ways, it took up so much of the field of vision that many in the crowd had the sensation that they were moving backward.

The ship splashed into the cold Thames River. The crowd cheered at another step in the march of nuclear progress, but in the way the *Nautilus* rode high in the ocean there was a visible sign of how much work remained to be done. Much of the ship's largest, heaviest equipment had yet to be installed, including the most crucial component of all: the nuclear reactor. The final assembly would take nearly all of 1954.

Finally, on December 30, 1954, the reactor was installed, tested, and ready to achieve criticality. As the start-up progressed, Theodore Rockwell, a civilian and one of Rickover's most trusted engineers, deftly manipulated his slide rule and plotted one of those seemingly obscure ratios that mean so much in reactor dynamics: the inches of rod withdrawal versus the inverse of the neutron counts per second. The precise math that governed everything in nuclear engineering said that such a graph would generate a straight line that intersected the horizontal axis at the exact point of criticality. Thus, after hours of meticulous, slow withdrawals and infinitesimal increases in neutron counts per second, Rockwell used the graph to predict at what height the control rods would make the reactor critical. It was a crucial concern, and everyone in the packed room followed Rockwell's graph intently. One potential problem in a brand-new reactor was that there might not be enough fuel loaded to achieve criticality, even with rods fully withdrawn—this had actually happened to Zinn and his EBR-1 back in Idaho. On the other extreme, a far more worrisome scenario had criticality arrive too quickly, causing power to increase exponentially before the instruments could even catch up with it, damaging the core or worse before any of the reactor's automatic protection systems could save it from itself. Because of that scenario, the reactor operator pulled the rods up in steps, allowing a series of readings to be taken as the rods rose incrementally toward critical height.

Finally, just before midnight, the reactor of the *Nautilus* became critical for the first time. Twenty-one months had passed since Rickover's reactor in Idaho had reached the same milestone.

After two weeks of testing, with Captain Dennis Wilkinson in command and Rickover always at hand, the ship was finally ready to go to sea for the first time. Wilkinson, an experienced submariner, insisted on a two-day "fast cruise," so called because the operation takes place with the ship tied "fast" to the pier. During the cruise, every piece of equipment was tested, and all of it was operated exclusively by the ship's crew. This included many of the systems that, while not exactly nuclear, would be crucial to the *Nautilus:* systems that purified the air during extended underwater cruising, distilling units that would make out of seawater the highly purified water necessary for operating the reactor plant, and even the refrigerators that would keep food stores safe. To maintain the simulation, no one on the crew was allowed to leave the ship for the duration of the fast cruise.

In the middle of this crucial, hectic operation, Wilkinson was annoyed to hear that two men from the office of the Navy's chief of information, or "Chinfo," were at the pier demanding to talk to him. Breaking with the simulation only slightly, Wilkinson went topside and shouted across to the public relations men on the pier, as the crew beneath his feet hurriedly completed the preparations for going to sea.

"What you're about to do is historic," they shouted to the captain. He needed to be ready to mark the occasion with a "historic message" upon getting under way.

Wilkinson, who had his hands quite full trying to get the submarine ready, responded, "You're the communications specialists. You write it."

The PR men walked off. They were happy to compose the message and Wilkinson was happy to see them go. The next day, they returned and proudly delivered their work. It was over a page long, filled with florid language about the dawn of the new age.

What the PR men didn't know or understand was that such a message would be communicated from the *Nautilus* not via the spoken word, or even teletype, but with the use of flashing lights and Morse code. And on a submarine, the signal lamps wouldn't even be operated by signalmen, but by cross-trained sailors who might be able to manage ten words a minute. Wilkinson kept the Chinfo message as a memento but never for a second considered transmitting it.

The next day, at 11:00 AM, on a cold gray morning, the USS *Nautilus* steamed away from the Groton pier and into the Atlantic Ocean. As the throttleman turned his wheel, steam flowed to the main engines, which, through a reduction gear, turned the screw in direct response to Wilkinson's skilled orders as he maneuvered the ship into the channel. Admiral Rickover stood quietly next to him. Here, the revolutionary form of power was thoroughly subordinate to a brass, bell-ringing engine order telegraph and the traditional orders of ship control: Ahead one-third . . . back two-thirds . . . all stop. There was a brief scare when a loose screw caused a racket in the main reduction gear, but the problem was quickly resolved, and overall the propulsion plant responded wonderfully to the rapid-fire engine orders required by a maneuvering submarine.

Once the ship was well into the Thames River, Captain Wilkinson remembered his obligation to transmit a message marking the historic occasion. Ignoring the announcement he'd been handed the day before by the Chinfo functionaries, he ordered his quartermaster to transmit via signal lamp a far more graceful declaration of the new, triumphant Navy: UNDERWAY ON NUCLEAR POWER.

It was January 17, 1955. Seven years earlier, before a nut had been turned or a single shovelful of ground had been turned over in Idaho, Rickover had predicted his nuclear submarine would get under way on January 1, 1955. He'd come within days of getting it exactly right.

THE RECOVERY

When Lamprecht and Moshberger hurriedly retreated down the steps of SL-1, they were greeted by a rapidly growing crowd in SL-1's gravel driveway. A "Class One" disaster had been declared, and as far away as Washington, D.C., important men started to wonder what the implications of the night might be. In Idaho, they were answering their phones, rubbing their eyes, and speeding in their government Studebakers toward the obscure broken reactor in the desert.

Among the first to arrive were Ed Vallario, the chief health physicist for SL-1, and Paul Duckworth, the overall supervisor of SL-1 from Combustion Engineering, the Connecticut-based contractor that managed the site for the Army. No one was sure if the three crewmen inside were dead or not. It was their job to find out.

They donned Scott air banks and hurried up the stairs, the radiation detectors confirming what Lamprecht and Moshberger had seen—the radiation levels were scorching. At the top of the steps, Vallario paused. The plant was silent, but over the sound of his own heavy breathing through the mask's respirator, he distinctly heard an agonized moan.

At his feet, motion caught his eye. The nearest body, the one Lamprecht had seen, was moving. The face was almost completely destroyed, and the body was soaking wet, but he was moving. To Vallario, it seemed like a primitive, instinctive motion, as if the man were unconsciously trying to distance himself from the source of danger, the reactor. The man had been brutalized by the explosion, the heat, and the radiation, but he was alive.

Vallario and Duckworth raced down the steps and recruited three more men to help move the survivor. By then, the men on the scene knew the names of the three men who were supposed to be on watch, and Vallario remembered meeting all three of them in the course of his work in months past. The survivor was so mutilated, however, that Vallario misidentified him as Byrnes. But it was McKinley, the luckless trainee who had just arrived in Idaho.

The crew rushed up the stairs and threw the body on a stretcher they grabbed along the way. Vallario automatically made a rough calculation of the radiation exposure as he went. He figured that twenty minutes in that kind of radiation should be lethal. The man they were recovering had been lying there for over an hour.

Inside the reactor building, they loaded McKinley onto a stretcher, and then carried him down the stairs as fast as they could. Both Duckworth and Vallario had their respirators fail while inside the building, and both men, after holding their breath as long as possible, were forced to remove their masks and take long, deep breaths inside the reactor building. Unlike Lamprecht, these men had both received extensive training in radiation health, and both knew at least some of the implications of breathing that contaminated air.

Outside, they shoved the body on its stretcher into a panel truck, which drove a short distance down the road, where it met the facility's Pontiac ambulance, purchased just the month before. In the ambulance rode Helen Leisen, the site's on-call nurse. She heard the battered man draw a ragged, painful breath as they sped away. She tried to fit a respirator over his shattered face.

Fillmore Avenue was the prosaic name given to the dead-end road that led north from Highway 20 to SL-1. At the intersection of Fillmore and Highway 20, a checkpoint had been hastily established. Monitors at the checkpoint were shocked to discover that even at the door of the ambulance, their instruments read 400 R/hour—the victim inside was that radioactive. Nurse Leisen hurriedly jumped out of the ambulance and in her place went the site's on-call doctor, John Spickard. The breath Nurse Leisen had heard was McKinley's last. Spickard declared him dead at 11:14 PM.

Now, for the first time, the authorities in Idaho had to figure out what to do with a highly radioactive corpse. Taking him to a morgue or a mortuary in Idaho Falls was out of the question. Besides being completely unequipped to deal with this kind of hazard, the drive would kill any ambulance driver.

Finally, it was decided that the driver would just drive the ambulance off the road and into the scrub about a half mile away, and remain parked there until a better plan was developed. The nervous people at the checkpoint watched as the ambulance bounced into the brush. Its lights went off, and then the driver jumped out of the cab and separated himself from the ambulance as fast as he could run. At one point, workers thought they might reduce the radioactivity by cutting off McKinley's clothing, which was heavily contaminated. Soaked by water and frozen by the Idaho winter, however, the workers found the body to be almost completely encased in radioactive ice, and the clothing was as solid as concrete. With great effort and heavy cutting tools, they finally managed to cut off McKinley's uniform, but found that it reduced the radioactivity only negligibly.

The second body, eventually found to be Byrnes, was removed from SL-1 in much the same way as McKinley, and in the same ambulance. Both bodies were taken to the Chemical Processing Plant and placed inside steel tanks filled with alcohol and ice, with the hope that it would preserve the bodies and reduce the contamination.

One of the many fables that would spring from the SL-1 disaster was that the brand-new Pontiac ambulance had to be buried after January 3 because it was so radioactive. Much of the debris from the accident, including substantial parts of the three crewmen, was buried nearby. However, while burying the ambulance would have been far cheaper, the long and difficult exercise was deemed a valuable and rare chance to practice the techniques of decontamination on a large scale. After an extensive cleaning, the ambulance was put back in service at the NRTS, serving in a variety of capacities around the vast facility for years.

The third body, Richard Legg, was discovered by a search party that entered SL-1 at 10:38 PM. The shaken men reported what they had seen pinned to the ceiling, a lifeless clump they initially thought was a bundle of rags.

Recovering Legg's body would be a tremendous challenge. It was inside a deadly radiation field, one in which rescue workers, even utilizing emergency limits, were only allowed to spend sixty seconds at a time. It had been hard enough to run in there with a stretcher and remove Byrnes and McKinley. No one understood how they could remove an impaled body thirteen feet up, over a gaping, smoldering nuclear reactor, without seriously endangering the rescue crews.

Secondly, poised over the reactor as it was, some engineers worried that if Legg's body fell inside the reactor during the recovery attempt it might start yet another unintended power excursion. The body or the rod that impaled it might be covered with enough fuel to cause criticality. Or, a falling body might just knock things around in the shattered reactor in such a way that a critical geometry was reached and SL-1 would run wild once again. No one had any idea what had happened at SL-1—they had been told over and over again that reactors were all "inherently safe." With so much confusion, and so many unknowns, no one wanted to take any chances. Teams were trained, a wooden mock-up was hurriedly built, and special crews were brought up to Idaho from Dugway Proving Ground in Utah to assist. Dugway was the site of the Army's new Chemical, Biological, and Radiological Weapons School, and the generals were eager to give their men some real-world experience with radioactivity.

Finally, after five days of planning, a team of volunteers was chosen to remove Legg's body on January 8, 1961.

A special boom had been constructed, and on the end of it a giant canvass stretcher, five feet by twenty feet long. This stretcher-boom was attached to the end of a crane, which could poke it into the reactor building through a service door designed to allow the entry and exit of large pieces of equipment. The crane was carefully driven up to the side of SL-1, and the boom was inserted through the door, until the stretcher was positioned directly beneath the impaled body of Richard Legg, directly above the reactor.

Once the net was in place, a team of ten soldiers, acting in teams of two, took 65-second turns inside the reactor building. Equipped with sharp steel hooks on the end of long poles, they took turns snagging and pulling at the dead flesh of Legg. The fourth team finally finished the job, and the body came free and crashed down onto the stretcher, at 2:37 AM on January 9, 1961.

The recovery team took a break to allow the exhausted crews to rest and to let the sun come up before finishing the job. The following afternoon, the crane slowly pulled the boom out of the building and deposited Legg's body into a cask specially made out of four-inch-thick lead panels on the back of a flatbed truck. The truck, with police escort, took its highly radioactive cargo to the Chemical Processing Plant, where Legg joined his two crewmates once again, this time in a stainless-steel room that no one dared enter.

chapter 4

THE ARMY

It is unlikely that the Russians were any more stunned by the rapid success of Hyman Rickover and the Navy than the other branches of the United States military were. After all, the three services had all started investigating nuclear power at roughly the same time. Rickover went to Oak Ridge in May 1946. That very same month, the Air Force signed a contract with Fairchild Engine and Airplane Corporation to conduct a feasibility study on the use of nuclear propulsion for aircraft. In some ways, the Army had the biggest head start of all, having been the overseer of the Manhattan Project at the dawn of the nuclear age. By 1955, however, Rickover's submarine was under way on nuclear power, while the other services were still mired in theoretical development and long-range planning. The Army scrambled to regain the initiative. At best, a significant chunk of the defense budget was at stake. At worst, the generals feared, if they could not find a way to turn the Army into a nuclear-powered force, they risked obsolescence.

At first, the Army was confident that nuclear weapons were the surest path to relevance in the atomic age. Nuclear weapons were tried, proven, and getting more powerful all the time, while nuclear power seemed a complicated fantasy—at least until the *Nautilus* was launched. In 1957, the Army's research and development budget for nuclear weapons was ten times greater than what it spent to research either artillery or aircraft, an illustration of the Army's devotion to weapons systems. Those research dollars soon provided the Army with a diverse nuclear arsenal.

Nuclear artillery shells were the first innovation. Nuclear artillery, the generals argued, could help equalize conventional force imbalances with the Soviet Union in Europe. Perhaps just as importantly, no rival service could co-opt artillery from the Army, the way the Air Force had with both bombs and the guided missiles that were then in the earliest stages of development. The nuclear artillery program began even before World War II ended, with the development of a massive 280-millimeter atomic cannon, one of which rolled down Pennsylvania Avenue in President Eisenhower's 1952 inaugural parade. The weapon was nicknamed "Atomic Annie," an homage to the giant (but conventional) German railway gun "Anzio Annie" that it resembled. While the monstrously large gun looked like some kind of medieval siege engine, it was built to fire the most modern of weapons: an eight-hundred-pound nuclear projectile a distance of around seventeen miles. The cannon weighed eighty-three tons and moved on the backs of two trucks, one on each end, as the two drivers communicated frantically with each other via a built-in telephone system.

The first atomic cannon was delivered in 1952, and was declared obsolete almost immediately after. Only twenty were ever manufactured. Critics, mostly from the Navy and Air Force, criticized the Army's gun as being clumsy and immobile, as well as an inefficient use of the nation's scarce fissionable material—this because the artillery shells used the same "gun" type configuration as Little Boy, rather than the more sophisticated implosion system used with Fat Man on Nagasaki. The Army fired a nuclear artillery shell only once, in a Nevada test code-named "Knothole Grable" on May 25, 1953. The shell exploded with a force of fifteen kilotons—about equal to the Hiroshima bomb—424 feet above the Nevada desert while thousands of Army troops looked on in wonder.

In contrast to the Navy and Air Force, which developed complicated command and control procedures that linked nuclear launch authority directly to the president, the Army distributed atomic firepower far down the chain of command. It developed smaller artillery shells, unguided missiles, and portable atomic munitions designed for commandos to use in destroying bridges and dams. The Army's miniaturization efforts reached their pinnacle with the "Davy Crockett," a nuclear warhead launched from a recoilless, jeep-mounted rifle with a four-man crew. The Davy Crockett had the distinction of being the smallest nuclear warhead ever manufactured by the United States. It weighed just fifty-one pounds and had a maximum yield of one kiloton.

But nuclear weapons, for all their increasing power and versatility, were not proving to be the cure-all for which the Army had hoped. The fighting in Korea ended in 1953, and despite some truly dire weeks for American forces during that conflict, the tactical and political situation on the peninsula made the use of even small atomic weapons impossible. Even before then, rumors of Rickover's progress were winding through the Pentagon, and the Army determined that it needed to develop some kind of nuclear power capability, just as Rickover was doing with his sub and the Air Force was doing with its atomic airplane. In 1952, the Army started a nuclear power program, and selected Colonel James B. Lampert to run it.

Born in 1914, in Washington, D.C., Lampert was the son of a father he would barely know: Lieutenant Colonel James G. B. Lampert was killed in Europe in January 1919 as part of the American Expeditionary Force. Before his death, the elder Lampert was in the Army Corps of Engineers, and was credited with having invented the floating footbridge. James Benjamin Lampert graduated from West Point seventeen years after his father's death, and was ranked 36th out of 276 cadets in his West Point class. (By that measure, he was superior to Rickover, 107th out of 540 in the Naval Academy's class of 1922.) West Point's Class of '36 would produce many of the leaders of the Vietnam era, men like Creighton Abrams and William Westmoreland, as well as Benjamin Oliver Davis Jr., who was a Tuskegee Airman and the first black Air Force general.

Like his father, Lampert joined the Corps of Engineers. During World War II he found himself at the epicenter of the engineers' most important endeavor—the Manhattan Project, where Lampert acted as executive officer for General Leslie Groves, the senior military commander on the project. By the time the Army decided to pursue its own independent nuclear power program in 1952, Lampert was, relatively speaking, richly experienced.

He was also in many ways the opposite of Rickover. While Rickover was the consummate outsider and held a grudge against the Naval Academy all his life, Lampert was a proud West Point graduate, the son of a West Point graduate, and husband to the daughter of a West Point professor and World War I hero Colonel William Mitchell. Lampert, all his life, cherished the U.S. Military Academy. Rickover once told Congress that the Naval Academy was "an aggregation of photographic memorizers." Lampert's tombstone would read "No One Loved West Point More." Also unlike Rickover, Lampert's career was not defined by nuclear power. He would become the first son of a West Point graduate ever to serve as the academy's superintendent, and he considered those years, 1963–66, not his leadership of the nuclear program, to be his most important tour of duty. Lampert also differed markedly from Rickover in the way he treated subordinates. One close associate who worked for Lampert for years would say of him later, "he didn't lose his temper," and that if he felt the need to criticize, "he always started out by complimenting me." It is not a portrait anyone would confuse with Rickover's.

Lampert served his tenure as head of the Army's nuclear program as a colonel—the Army's equivalent of a Navy captain, the rank Rickover had fought so hard to surpass. Lampert would make general in 1958, after departing the nuclear program, and was eventually promoted to the rank of three-star general. That's the rank he held during his last tour of duty, as the last U.S. High Commissioner of the Ryukyu Islands in Japan from 1968 to 1972.

One characteristic of Rickover's the Army did want to emulate, however, was his bicameral job description, with one foot in the military and the other in the Atomic Energy Commission, an arrangement everyone had seen Rickover deftly manipulate to his advantage. The Atomic Energy Commission, perhaps starting to see the monster they had helped create in Rickover, at first resisted, but in the end they couldn't grant that status to a Navy admiral and then deny it to an Army general.

Lampert early on in his tenure actually sought out the then-captain Rickover, to introduce himself and seek advice. Rickover, in the thick of his promotion fight and the struggle to get the *Nautilus* constructed, gave the colonel a predictably gruff introduction to naval nuclear power. "I understand you want to build nuclear power plants for the Army," said Rickover. "My advice to you is that you don't know what you're doing and the best thing you can do is get out of it in a hurry." Rickover did, in the end, grudgingly give Lampert access to the Navy's rapidly accumulating library of research and technical reports.

From the beginning, Colonel Lampert and his colleagues had to answer a fundamental question: Exactly why does the Army need nuclear power plants? The simple answer was that a small, mobile power plant that rarely required refueling would have real advantages for the Army in remote locations, such as Greenland and Alaska. Fuel was delivered to those bases almost exclusively via airlift, a difficult enough endeavor in peacetime. Few expected these remote stations to stay energized if World War III ever broke out. Still, to some, it hardly seemed worthwhile to begin the massive, costly enterprise of reactor development just for the sake of a few, remote bases in theoretical jeopardy.

The situation changed dramatically on February 15, 1954, the day President Eisenhower announced one of the most ambitious civil engineering projects of the twentieth century: the Distant Early Warning Line. The DEW line would be a massive chain of radar stations two hundred miles above the Arctic Circle, an impregnable electronic shield through which no Soviet bomber could fly undetected. Interestingly, the DEW line was born at almost the same moment Eisenhower was initiating another impossibly big public works project: the interstate highway system. American presidents had for decades tried unsuccessfully to build a national network of highways. Eisenhower himself had seen the need during a slogging cross-country Army caravan in 1919. It took the specter of World War III, however, to finally mobilize the public will. Good highways, Ike explained, like the ones he had seen in Germany during the war, were needed to move American troops and materiel rapidly from city to city. Vice President Richard Nixon, in promoting the plan in a 1954 speech to the Governors Conference, stated the nation needed modern highways "to meet the demands of catastrophe or defense, should an atomic war come." In his 1961 farewell address, Eisenhower would famously warn Americans about the dangers of the "military-industrial complex." Part of his concern must have originated during those frantic months when the federal government poured tons of concrete and billions of dollars into these two massive Cold War projects. While Eisenhower's highways continue to affect Americans on a daily basis, however, the DEW line, like the atomic cannon, was declared obsolete practically the day it was completed.

The DEW line had originated with a study group at the Massachusetts Institute of Technology, the nation's academic incubator for many of its most grandiose Cold War schemes. The university's Summer Study Group of 1952 warned that America was dangerously vulnerable to an air attack, especially one flying a polar route, the shortest route for Soviet bombers. The United States did already have under construction two parallel lines of radar stations to guard its northern flank. The southernmost line, the "Pinetree Line," roughly followed the U.S.-Canadian border. The second line, being completed as the DEW line began, ran at about the 59th parallel and was designated the "Mid-Canada Line." As Soviet bombers grew faster and more sophisticated, however, more warning was required, pushing the required radar pickets farther and farther north. A line at the 69th parallel, two hundred miles inside the Arctic Circle, would roughly trace the northernmost edge of the North American landmass. State-of-the-art radar stations there, the MIT scientists theorized, would provide America with something like a three-hour forewarning of a Soviet air attack.

The huge primary construction contract for the line was awarded to Western Electric, the manufacturing arm of AT&T, in December 1954. The target date for completion was July 31, 1957. It was a timetable every bit as ambitious as Rickover's for completing the *Nautilus.* And just like Rickover's submarine, the project was completed precisely on time. In a span that included just two short Arctic summers, Western Electric and their contractors completed in the most remote locations on earth, in the most severe conditions, an interlocking chain of over fifty radar stations, some unmanned and automated to an unprecedented degree, some fully manned, self-contained communities.

The construction of the DEW line required the largest civilian airlift in the history of the world. In all, over a half-million tons of material were moved from the United States to the stations. Western Electric congratulated itself in a 1960 brochure about the project, and cited some of its more jaw-dropping statistics, most of which revolved around the sheer tonnage transported and manufactured north of the Arctic Circle: enough gravel to build two Great Pyramids, twelve acres of bedsheets, three miles of window shades.

The statistic that most interested Lampert was the staggering amount of fuel required to run the DEW line. Total generating capacity of the system was 155,000 kilowatts: about the same amount of energy required by the city of Spokane, Washington. Just during construction, seventy-five million gallons of petroleum were transported to the line, "enough to fill the tank cars of a train 65 miles long." Most of this fuel was transported laboriously across the permafrost in steel barrels. And unlike bedsheets or window shades, oil would be constantly consumed by the radar stations as they rotated their antennas and tried to heat their Spartan barracks. Keeping the DEW line supplied was difficult enough during peacetime. In a shooting war, no one thought the gossamer-thin supply lines could keep the stations supplied for long. Their radar screens would go dark, their crews would slowly freeze, and America would be once again blind to the attacking jets of the godless Soviets.

Nuclear power was an obvious solution. One power plant could keep a base powered for years without refueling. The DEW line gave the Corps of Engineers and Colonel Lampert the strategic imperative they'd been looking for: they would create small, semiportable nuclear power stations capable of powering remote Arctic bases. While Lampert might not have been as daunting a presence as Rickover, he took his mission every bit as seriously. As one contemporary account of the DEW line construction stated, Lampert and his colleagues believed that "the success of their project might mean the survival of the nation." They swung into work.

For his initial reactor design, Lampert decided to follow in large part the successful Navy model: a pressurized reactor that used water as both coolant and moderator. The decision to make the reactor pressurized was a significant one. In a boiling water reactor, water is boiled directly in the core, and that steam was used to turn the turbines. This eliminated entirely the need for heat exchangers and the second "loop" required for a pressurized reactor, a major concern in a plant that was meant to be modular and portable. Rickover had contemplated the same choice—after all, space is also at a premium inside a submarine. In the end, Rickover felt the inherent safety benefits of the pressurized design made it worth the extra weight and space. Lampert and his team reached the same conclusion.

While following in many ways Rickover's successful model, Lampert did have requirements that were unique to the Army. The reactor had to be modular, with no single piece being larger or heavier than what a cargo plane could carry to an Arctic outpost: seven by seven by eighteen feet, with no single piece heavier than ten tons, except for the generator and the main condensers, which were allowed to weigh up to twenty tons each. Additionally, erection of the entire plant on-site could take no more then six months, a requirement designed with the short Arctic construction season in mind. Finally, the plant had to generate 1,000 kilowatts of electricity, and, in addition, enough steam to heat an Arctic garrison of up to two hundred men.

The Army departed from the Navy in one other significant aspect of its nuclear program. The Army decided early on to make the program, to the maximum extent possible, unclassified. The idea was in keeping with Eisenhower's Atoms for Peace ideals, to make the wonders of nuclear energy available to all. Or, in the words of Army Secretary Frank Pace, "the far reaching benefits of this program in terms of peaceful application should be made known as fully and rapidly as possible to the American public." It was a major departure from the Navy's philosophy.

At first glance, Rickover appeared anything but secretive. He managed an adoring corps of reporters and had a genuine knack for staging imaginative publicity events. For example, even as the Army program was getting off the ground, Rickover hosted a meeting for the Joint Committee on Atomic Energy aboard the Nautilus, entertaining his powerful guests by quipping that they were now "a sub-committee" as the ship submerged, and as the Nautilus took an angle that he was delivering "slanted testimony." On February 4, 1957, the Nautilus marked 60,000 miles under way. Or, as Rickover's press releases adroitly pointed out, 20,000 leagues. A commemorative telegram from the ship's sponsor, Mamie Eisenhower, marked the whimsical milestone. Nevertheless, despite Rickover's love of good publicity, all but the most cursory technical details of the S1W reactor were kept tightly secret. Even today, the exact power capacity of the S1W reactor remains classified, and the S1W prototype in Idaho, decommissioned since 1989, is one of the few historic locations at the site that remains strictly off-limits to visitors.

The Army's choice of locations for its prototype plant mirrored its desire to build the program squarely in the public eye. Rather than in the wilds of Idaho, the Army would build its first nuclear power plant at Fort Belvoir, Virginia, the headquarters of the Army Corps of Engineers. It is one sign of how the perception of nuclear power has changed, the decision to build an experimental nuclear power plant in that densely populated area, just eighteen miles from the White House. In December 1954, the Army awarded the $2 million contract to build the plant to ALCO Products, a venerable Schenectady, New York, manufacturer formerly known as American Locomotive.

The plant was initially called APPR-1, for Army Package Power Reactor #1. The name changed in 1958 when the Army created a standard nomenclature, the breadth of which indicated their optimism about their nuclear future. Plants would be designated with a first letter that indicated whether it was stationary (S), mobile (M), or portable (P). The second letter indicated power level: high (H), medium (M), or low (L). The final number indicated how many plants of that type had been built. Thus the Army's prototype at Fort Belvoir was designated SM-1.

Construction began on October 5, 1955, and proceeded smoothly under the skillful management of ALCO, Lampert, and his team. Despite the generally friendly mood toward nuclear power in that era, some civilian neighbors did fret about the reactor, one complaining that radiation was killing her roses long before the core was even fueled. In response, an Army major with the program planted roses inside the fences of SM-1 to prove her wrong, and at least one of those bushes, according to program historian Lawrence Suid, was still in bloom twenty-six years later. The plant went critical for the first time on April 8, 1957, just eighteen months after the start of construction. It was formally dedicated three weeks later. In the ceremony, SM-1's electricity was used to power both a radar antenna and a printing press, bluntly symbolizing the civilian and military potential of nuclear power. Lampert was disappointed that President Eisenhower declined to attend the ceremony, but he could take satisfaction in the fact that his power plant had gone critical eight months earlier than Rickover's "civilian" power plant in Shippingport, Pennsylvania.

The primary mission of SM-1 was to train operators. Byrnes, Legg, and their classmates would study the principles of nuclear theory in an eight-week crash course inside a classroom before moving over to the actual plant for hands-on instruction. They would also learn about the mysterious effects of radiation on the body, "health physics," and that even the mortal dose of radiation to a human was not a straightforward datum. The key number was called "LD 50-30," shorthand for the dose that would be lethal to 50 percent of the population within thirty days. Exactly how a large dose of radiation killed people was relatively straightforward—among other things, it destroyed the ability of the blood supply to renew itself. More mysterious were the effects of lower doses of radiation on genetics and reproduction: a contemporary book on health physics likened the effects of radiation on chromosomes to "the snapping of a cable by a rifle bullet."

Exposure to radiation is reduced by three, and only three, factors:

time, distance, and shielding. Cut your time in a given radiation field by half, and you receive half the dose. Pre-staging tools and rehearsing procedures are time honored ways to accomplish this. Shielding, putting something between you and the radiation source, also reduces radiation exposure—heavy, dense materials provide better shielding than thin ones. A two-inch panel of solid lead, one of the best shielding materials, for example, reduces gamma radiation by about a factor of 10. Distance is one of the most effective methods for reducing radiation, as the dose rate from a point source of radiation drops in relation to the square of the distance. In other words, if a radiation level is 100 R/hour at one foot distance, it will drop to 1 R/hour by just moving ten feet away. None of these methods will reduce radiation to zero, but some combination of the three can keep the radiation dose "ALARA": As Low as Reasonably Achievable.

It is typical of nuclear science that three different units would be developed to measure the same thing, are in many cases equal, and are all at times abbreviated with the letter "R." The men at Fort Belvoir learned about "roentgens," a venerable unit named after German physicist Wilhelm Roentgen. The actual amount of actual biological damage caused by a given amount of radiation, or "dose," differed, depending on how the exposure occurred and what part of the body was exposed. The Roentgen was multiplied by a quality factor to account for these differences, and was called "Roentgen Equivalent Man," or "rem." For gamma rays or X-rays, one rem equaled one roentgen. Both these units superseded an older unit, the "Radiation Absorbed Dose," or "rad." All of these have since been superseded by two metric units, the gray and the sievert, neither of which, thankfully, begins with the letter *R*.

The training course at SM-1 was not designed to create theoretical physicists. It was designed to take some of the sharpest soldiers in the Army and make efficient, safe reactor operators out of them. Complicated concepts were simplified, at times with an elegance and a precision that Fermi himself would have admired. The difference between contamination and radiation can be confusing to the uninitiated: the former is a physical substance that can be moved around, the latter is energy that can be attenuated but not put in one's pocket. "Radiation is stink," generations of military nukes have learned, "contamination is shit."

Slightly more than three months after SM-1 went critical, on July 31, 1957, the DEW line was declared complete. The line, a marvel of technology, made up a twelve-mile-high, 3,000-mile-long radar fence along the northern edge of the continent. A ceremony took place near the western terminus of the line on August 13, at Point Barrow, Alaska, during which Western Electric executives handed over control of the line to the military. The generals commended the contractors, thanking them for completing a project "of utmost significance to the defense of the North American continent." The United States would now have approximately three hours' forewarning of Soviet bombers flying over the North Pole. With plants like SM-1 powering the DEW line, the radar stations would become even more reliable sentries for democracy.

For all its wonder, however, the DEW line was a fixed defense. And from the Great Wall of China to the Maginot Line, history loves to show military planners the folly of fixed defenses. (Some perceptive critics had during its construction called the DEW line the "imaginot line.") For the DEW line, the end would come extraordinarily fast. The project cost a billion dollars, twenty lives, and took thirty-two months to build, but as a viable defense it would succeed for only nine weeks.

On October 4, 1957, the Soviets launched Sputnik.

Sputnik was the Soviet *Nautilus:* a singular, decisive technological victory that left experts on the other side of the Iron Curtain gasping in surprise. The United States had itself confidently announced a satellite program much earlier, Project Vanguard, with the stated goal of putting a satellite into orbit during the International Geophysical Year. It was during an IGY conference in Washington, D.C., that the Soviets announced the successful launch of Sputnik. The Soviets were straightforward in their description of the satellite; it was such a self-evident triumph that the normal need for propagandizing was absent. The satellite itself was not much of a military threat, a 184-pound, beach ball–sized sphere orbiting the earth at 18,000 miles per hour. It contained only radio transmitters and batteries, but the thought of it circling above the United States was incredibly unsettling to that large group who had presumed the West was technologically superior, and that the Soviets' rare technological victories, like their atom bomb, were the result of espionage rather than scientific prowess. Sputnik proved that theory spectacularly wrong. *Time* summed up the somber mood with its headline: "Red Moon over the U.S."

While the steady beeping of Sputnik, audible to ham radio operators all over the world at 20 to 40 megahertz, may have been the sound of a propaganda triumph, it was the rocket that launched Sputnik that was most worrisome to the Pentagon. The Soviets had constructed a viable ballistic missile—they had, in this case, used it to push a satellite into space. But no one doubted that the same missile could be used to lob nuclear weapons onto a defenseless United States. The DEW line couldn't provide the slightest protection. The high, parabolic trajectory of a long-range missile would take it far above the radar net. General Earle E. Partridge, the commander of the North American Air Defense Command, or NORAD, sounded stunned when he told a reporter shortly after the launch of Sputnik, "if the aggressor's weapon is the ICBM, the continent stands today almost as naked as it did in 1946, for I have no radar to detect missiles and no defense against them."

• • •

Despite the devastating effect of Sputnik on both the American psyche and the value of the DEW line, the Army pressed on with its nuclear power program. The success of SM-1 at Fort Belvoir had inspired some in the Army to fantasize about more exotic uses for nuclear power. And just as with nuclear weapons, the Army's inclination was to reduce reactors in size, to make them ever more portable, and to push them further and further down the chain of command. Colonel William Gribble, Jr., one of Lampert's colleagues, briefed the Army's Transportation Corps on March 15, 1955, telling them that a nuclear-powered snow train (presumably to haul nuclear power plant components to remote DEW line stations) was "quite feasible." A subsequent design study concluded such a train would have up to ten cars, each with a thirty-ton capacity, rolling on wheels ten feet in diameter. The Army also speculated about using nuclear power for locomotives, large trucks, and even a nuclear-powered tank, a behemoth that would weight at least fifty tons. The Army did allow that the destruction of such a tank on the battlefield could create "a low order atomic explosion." It was a persistent part of the atomic dream, these compact nuclear engines that would power machines of all sizes.

While they were not as glamorous as a nuclear-powered snow train with ten-foot wheels, the Army devoted most its energy to following up the success of SM-1 with more small power plants. After all, the DEW line was still in operation, and whatever new antimissile systems were eventually developed in the post-Sputnik scramble, they would probably utilize at least some elements of the DEW line. Small, modular, nuclear power stations would still be militarily valuable. The Army began designing its next prototype, a plant that would be even smaller than the one operating successfully at Fort Belvoir. Rather than generating one or two thousand kilowatts of electricity, enough for a small base, this new plant would generate just two or three hundred: enough for a small radar station and crew. They would call it SL-1, and this time, the prototype would be built in Idaho.

Lampert, emboldened by his success at Fort Belvoir, decided to depart from the Navy-style pressurized reactors, which generated steam in a secondary, nonradioactive loop. For SL-1, the Army would construct a boiling water reactor, the kind of reactor in which the core boils water directly and in effect serves as its own steam generator. The design saved considerably on the size and the amount of equipment necessary to run the plant, as the entire, elaborate heat-exchanging apparatus was eliminated.

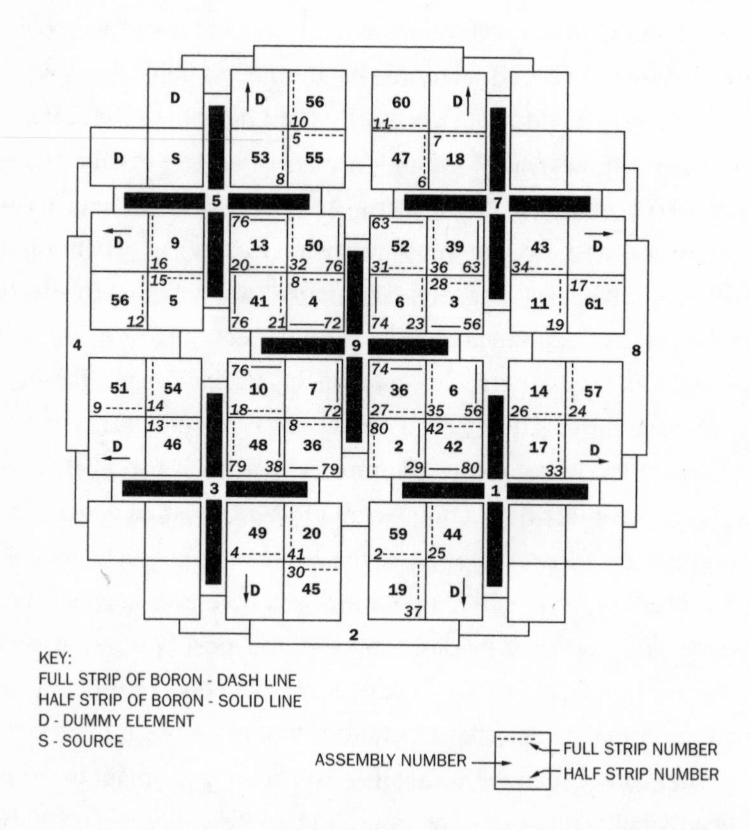

The positions of fuel, boron, and control rods in the SL-1 core.
The five cross-shaped control rods are represented by the bold lines.
(From 19300, the interim report.)

In return, boiling water reactors (BWRs) were in some ways inherently less stable than pressurized reactors. Boiling water reactors created steam inside the core, while pressurized reactors kept the core covered with liquid, and thus cooled, at all times. Additionally, the steam turning the generating turbines for a BWR was radioactive, greatly complicating maintenance and operations. In a pressurized system, such as in the *Nautilus*, all the components of the engine room were more or less conventional and nonradioactive.

Finally, again in the interest of simplicity and reducing size, SL-1 was built with just five control rods, arranged like the dots of a five on a die (SM-1 in Fort Belvoir had seven rods). The rods were not numbered simply 1 through 5, but were instead designated by their relative position among the fuel assemblies: 1, 3, 5, 7, and the central rod, 9. That rod, by virtue of its position, would be enormously powerful. It would by itself contain enough reactivity to shut down the core if shoved to the bottom, or to start it up if pulled to the top. To build that much power into a single control rod fundamentally reduced the margin for safety inherent in the reactor. Most reactor designs adhered to the "one stuck rod" criteria, which held that no single out-of-control rod, even if fully withdrawn from the core, could push the reactor to criticality.

The prototype in Idaho would adhere to its Arctic simulation in several ways. It would be built without a containment building, seen as unnecessary because of the low power level and the relatively unpopulated area both in Idaho and in the plant's potential Arctic locations. Because permafrost couldn't support a traditionally constructed building, the reactor and its associated buildings would be built on concrete pilings. Gravel would be used for radiation shielding, as it was one of the few manufacturing materials that could be constructed on location in the Arctic. The maximum allowable size for components had grown slightly since SM-1, reflecting the Air Force's growing cargo planes: eight by nine by twenty feet.

The Army again departed from its successful experience with the SM-1 reactor in its choice of contractors to build SL-1. Because of their experience in the relatively new field of boiling water reactors, the Army hired Argonne National Laboratory to design and build the reactor. The builder of SM-1 at Fort Belvoir had been ALCO, a manufacturing giant that had been building products for industry for over fifty years. Like Rickover's vendors of choice, Westinghouse and GE, ALCO's expertise was in taking complicated scientific principles and bringing them to life in a rugged, operational form that was practical and reliable in the real world. Argonne was a laboratory, an entity whose specialty was the design of experimental systems for and by scientists. It was exactly this clash, the tension between scientists and engineers, that had made relations so frosty between Rickover and Walter Zinn, Argonne's first director, in the early days of the Navy program. SL-1 was meant to be a plant operated by a small cadre of young soldiers in remote areas. This kind of practicality was not a specialty of Argonne National Laboratory.

Ground was broken on SL-1 in the fall of 1956. The plant went critical on August 11, 1958. Soon after, the Army signed a contract with Combustion Engineering to manage the day-to-day operations of the plant. The Army had for some time been training future SL-1 operators at SM-1 in Fort Belvoir, in rigorous eight-month sessions. The Navy and Air Force had remote bases of their own, and were interested in the Army's small, modular plants, so they also sent men to be trained. In the Army's first class at Fort Belvoir for future SL-1 crewmen, twelve students were in the Army, three were from the Navy, and four were from the Air Force. Soon a steady trickle of soldiers and sailors were making the long trip from Virginia to Idaho. The fourth class of SL-1 operators reported to Idaho in the fall of 1959. That class included Richard Legg and Jack Byrnes.

• • •

By the time Byrnes and Legg reported to SL-1, the National Reactor Testing Station was starting to live up to its grandiose name. Ten years after its founding, the site was home to a dizzying variety of reactors, twenty-nine in all, each with a colorful name and a mission to test some new facet of the fledgling nuclear science. The BORAX reactors, their name an abbreviation of "Boiling Water Reactor Experiment," were built to test the limits of reactor safety and control. In a foreshadowing of SL-1, BORAX-1 was deliberately destroyed in a spectacular 1954 experiment in which a control rod was withdrawn until the reactor exploded, sending a radioactive plume of steam and debris one hundred feet into the sky. The SPERT reactors ("Special Power Excursion Reactor Tests") were also designed to test a reactor under runaway conditions, under a variety of hazardous temperature, pressure, and flow conditions. No theory in Idaho was too strange to be tested. The Organic Moderated Reactor Experiment used a liquid hydrocarbon as a coolant, rather than water or liquid sodium. Polyphenyl, a floor-wax-like substance, was noncorrosive and had an attractively high boiling point that the scientists couldn't resist trying. They learned to their disappointment that inside a reactor, polyphenyl turned into sticky tar. Among these engineering and scientific swans in Idaho, SL-1, designed to generate merely 200 kilowatts of electricity, was an ugly duckling. To make matters worse, as Legg and Byrnes reported in 1959, the small reactor seemed to be falling apart.

Almost from the beginning, SL-1's control rods had been sticking, a fundamental, dangerous problem. Control rods were the main mechanism for controlling reactor power, and the failure of a control rod to move on command represented a severe deficiency. Additionally, the sticking control rods indicated that something was mechanically amiss inside the core, blocking the rods or binding them inside the fuel channels in which they were meant to slide freely up and down.

In all, between February 1959 and December 1960, the five control rods inside SL-1 stuck sixty-three separate times. The powerful central control rod, rod 9, had malfunctioned seven separate times, dropping too slowly once, sticking in its channel during a scram five times, and once sticking so badly that even the drive motor could not move it. Even so, rod 9 behaved better than any other rod, probably because as the powerful, central rod, it was moved more frequently, which seemed to forestall stickiness. While it may have been the least defective rod, however, it was far from reliable. It is telling that many future accounts of SL-1 would say otherwise.

Idaho Falls, an account of the accident by William McKeown, would state that the central rod operated flawlessly before the accident: "the central control rod—the only rod that could have caused such devastation—was the only rod that didn't have a history of sticking. It had always slid in and out of the reactor with ease, just as it should have." The interim report written by Combustion Engineering for the Atomic Energy Commission immediately after the incident would also get it wrong. The report stated in its first few pages that the "No. 9 rod has the best over-all operational record and had successfully scrammed 130 times during the six months prior to the last shutdown period, with only one instance of sticking where it hesitated momentarily at the start of a scram." That same report, in its appendices, however, gives the actual data about rod 9, and details each of the seven sticking incidents. The misimpression that rod 9 was completely dependable was indicative of the strong reluctance of men closest to the program to find any fault with their machinery, which led to their equally strong inclination to blame a renegade operator.

In response to the troubling issue of the sticking rods, the Army ordered each shift of workers to "exercise" the rods: to regularly move the rods up and down to keep them from seizing. It wasn't working. On December 19, 1960, just fifteen days before the accident, two of the rods were stuck so severely that a pipe wrench was used to move them. Much of the Army's maintenance on SL-1 had that kind of makeshift feel to it. This was due, perhaps, to a general feeling that nuclear power was becoming routine. It was also, perhaps, a by-product of the Army's culture, a culture that prized ingenuity and improvisation. While doing whatever it takes to keep a tank operating on the battlefield is commendable, however, makeshift measures were questionable in a nuclear reactor. A 1960 quarterly report, for example, details how some enterprising soldiers fixed a leaky electrical connection on SL-1's purification pump by plugging the hole with an automotive spark plug.

Colonel Lampert had left the nuclear program in August 1957, moving on to a tour at the National War College and a promotion to brigadier general. Later would come his tours as superintendent of West Point, then High Commissioner of the Ryukyu Islands, each accompanied by a commensurate promotion. It was the model of a traditionally successful military career, a steady, successful progression through a wide variety of billets, culminating with a general's stars. It was exactly the type of career Rickover had before nuclear power, and one that he successfully battled to escape, so that he could do the opposite—remain in the same job for decades, building expertise and ensuring continuity as the naval nuclear power program grew in his image. Lampert had grown the Army program from scratch during his five years in command. One wonders how the situation would have been different had he stayed in that job beyond August 1957, a full year before SL-1 was even critical. Lampert was replaced in the position by Colonel Donald Williams, yet another alumnus of the Manhattan Project, a man who had spent just six months with the Army Reactors Branch when he took command of the program.

Williams and the rest of the Army leaders were seemingly unconcerned with identifying the root cause of the sticky control rods at SL-1. The men operating the plant, however, were sure they knew. SL-1, like many nuclear plants, had been constructed with "burnable poison" distributed throughout the core. The poison was a material that, like the control rods, had a very high affinity for neutrons and thus inhibited the nuclear reaction. Putting poison in the core helped balance the abundance of fuel early in core life. As fuel was consumed, so was the poison, and the theory was that this would help lengthen core life and keep the critical control rod height relatively constant throughout. In the case of SL-1, the burnable poison was boron, and strips of it had been tack-welded inside the core adjacent to the fuel elements.

From the start, the welds failed and boron flaked off into the core. It was first noticed during an inspection of the fuel during an August 1959 shutdown, just a year after the plant began operating, when the boron strips were observed to be bowing in the three-inch intervals between tack welds. By 1960, large amounts of boron were completely missing from the strips and the fuel elements, especially the crucial central elements, were difficult to remove because of the flaked-off boron clogging their channels. When the fuel elements were finally forced upward for inspection, uranium plates and boron fell off in chunks. The Army's response to the boron problem was symptomatic. Rather than isolate the reasons the interior of the core seemed to be disintegrating, and analyze the consequences, they simply ordered the cessation of the inspections.

The flaking boron had another sinister effect on the dynamics of the SL-1 core, beyond the obstruction of rod movement. As the poison flaked off and fell to the bottom of the core, it essentially increased the effective power of the core—removing poison from the most active region of the core added "reactivity," making the control rods relatively less powerful and the core perpetually closer to criticality, less controllable, and prone to unpredictable spikes in power.

A historic episode during the construction of the *Nautilus* shows how Rickover reacted to a similar problem in the Navy program. On September 16, 1954, testing was under way in the engine room of the *Nautilus*. As the reactor was not quite complete, steam was piped into the engine room from a boiler on the pier, a common practice in the shipyard. Shortly before midnight, a small pipe broke inside the hull, filling the engine room with steam. Casualty procedures were immediately followed, the steam was diverted to shore, and no one was injured. A crisis, it seemed, had been averted.

But Rickover was livid. He ordered an immediate and rigorous examination of the problem. It was discovered that the pipe that burst was not the high-quality seamless pipe it was supposed to be. It was instead rolled and welded, the kind of low-quality tubing used to construct handrails. Rickover further found that the records for this piping were not sufficiently detailed for them to ascertain exactly where else the rolled pipe might have been installed in error, either on the *Nautilus* or on the S1-W prototype in Idaho. Thousands of feet of the small diameter pipe had been installed on both plants, and all of it was covered and insulated, preventing a visual inspection.

Although the accident had caused no injuries and had taken place in a part of the plant that was ostensibly nonnuclear, Rickover ordered every inch of the pipe ripped out and replaced, a costly and time-consuming repair that to many seemed like a typical Rickover overreaction. Furthermore, Rickover mandated a detailed new quality assurance system be put in place, to prevent similar breakdowns and to create the kind of meticulous records necessary to track down similar problems in the future.

The Navy and the Army looked at nuclear power in fundamentally different ways. The Army, as indicated by its plans for nuclear tanks and locomotives, believed that nuclear power was just a wonderful new form of energy, a natural step in the same evolution that gave rise to coal-fired boilers and diesel engines. They treated their reactors and their nuclear-trained personnel accordingly. They were special, perhaps even elite, but they were not fundamentally different.

Rickover saw nuclear power as "something new under the sun," as Lewis Strauss said during the christening of the *Nautilus*. The admiral therefore created a program, a ship, and a corps of leaders who were also fundamentally different. While Rickover was a supremely skilled propagandist for himself and for the nuclear Navy, he was no spokesman for the concept of nuclear power as a panacea. After the *Nautilus* visited New York Harbor in August 1958, he quietly banned nuclear vessels from visiting large cities, a ban that lasted for decades. Despite the perfect safety record of his ships, he thought it too risky. He often spoke publicly about nuclear power as a necessary evil, something that required caretakers of extraordinary diligence and dedication. In a chance meeting on a train in 1954, the first chairman of the Atomic Energy Commission, David Lilienthal, suggested to Rickover that the utopian optimism swirling around nuclear power was unwarranted. Lilienthal recalled Rickover's reaction: "To my surprise, instead of rearing back and letting me have it—as I expected and almost counted on—his little face grew very sad. He couldn't agree with me more; why do people say things that don't make sense, and mislead people?" The Father of the Nuclear Navy summed up his mixed feelings about nuclear power again, shortly before the completion of the civilian nuclear power plant at Shippingport: "The whole reactor game hangs on a much more slender thread than most people are aware. There are a lot of things that can go wrong and it requires eternal vigilance."

The Army, in contrast, seemed to have accepted wholeheartedly that nuclear power was a benevolent, powerful ally to the American dream, dangerous in the same way automobiles were dangerous at the speeds they could attain on Eisenhower's new interstate highways: it was an entirely acceptable risk more than compensated for by the benefits of the new technology. By December 1960, procedures had been casually altered to account for the Army reactor's many problems: move the rods around every four hours or so, and stop the inspections that knocked more boron off their bowing strips. SL-1 kept generating power, the crewmen kept muttering under their breath about the troublesome reactor, and the days kept getting shorter and shorter as the darkest part of the Idaho winter took hold. A small respite was granted for the exhausted crews as the plant was shut down for the holidays.

On December 23, 1960, SL-1 was shut down by intentionally scramming the reactor. Of the five control rods, only two dropped cleanly to the bottom of the core: #5 and #9, the central rod. The other three rods had to be driven to the bottom of the core by their drive motors, what the interim report on the disaster would euphemistically call a "power assist." This benign terminology obscured the fact that the ability of a rod to drop cleanly to the bottom of the core was a key safety feature of any reactor. If everything failed, if all power were lost, gravity was supposed to pull the rods to the bottom of the core and safely shut down the reactor. It was at the heart of what made the reactor "inherently safe," in the argot of nuclear power proponents. That the rods could not travel to the bottom of the core without an assist was bad enough. But to make matters worse, the most likely culprit was crumbling boron strips, which put the reactor that much closer to criticality. It was an alarming condition to all who knew the plant, but it was no longer surprising by December 23. The rods had been sticking almost since the beginning at SL-1, and in the last days of 1960 the problem had gotten markedly worse.

A variety of maintenance was conducted on the plant during the Christmas shutdown. Instruments were calibrated, a condensate pump was overhauled, and a new type of valve was added to the auxiliary steam system. The biggest, most exotic maintenance performed on the plant was the addition of cobalt-aluminum wire segments throughout the core, which could be withdrawn later and used to measure the neutron flux patterns throughout the core. This maintenance was complicated enough that Combustion Engineering personnel did most of the work. To install the wires, the interior of the SL-1 core needed to be accessed. To do this, the control rod drive mechanisms had to be dismantled.

After the too-brief holiday break, and after the completion of the maintenance including the installation of the flux wires, reactor start-up was scheduled for January 3, 1961. All shifts were busy that day, refilling the reactor with water, replacing shielding, and performing the hundreds of checks and valve lineups required to get the plant in an operating condition. The start-up would occur on the night shift, the shift that would stand watch from 4:00 PM until midnight. Before they could raise the troublesome rods to restart the reactor, however, a long list of other tasks required completion, including the reassembly of the control rods. The shift tasked with this challenging maintenance was manned by Richard Legg, Jack Byrnes, and Richard McKinley.

It is often amazing what the military requires of young people, and often just as amazing what these young people are capable of doing. That being said, the plant superintendent's list of tasks for Legg and his crew on January 3 was extraordinary, especially when considering that the total SL-1 experience between the three men was less than three years. As was entirely normal, there were no officers present at SL-1. In Rickover's fleet, not only was an officer always present in the control room, he was an officer personally interviewed and vetted by the admiral himself. Without any additional supervision, and with only those few months of experience and some very sketchy written procedures to guide them, Legg and his crew were ordered to accomplish the following during their eight-hour watch on January 3, 1961, as listed concisely in the plant superintendent's night order book:

1. *Perform a reactor pump down—procedure No. 54*
2. *Reassemble the control rods, install plugs, place shield blocks, leave top shield off.*

3. *Connect rod drive motors.*
4. *Electrically and mechanically zero control rods.*
5. *Accomplish control room and plant startup check lists.*
6. *Perform cold rod drops.*
7. *At 300 psi pressure check for leaks, replace top shield plug.*
8. *Perform hot rod drop tests.*
9. *Accomplish a normal startup to 3MW operation.*

One can imagine Legg struggling to check items off the list, battling his own temper and the limitations of his only crew members, a brand-new trainee and a distracted, recalcitrant Jack Byrnes. Twice during the night, the work was interrupted by false fire alarms from the furnace room, when one of the men would have to drop what he was doing and meet the firemen at the gate to assure them there was nothing wrong at SL-1. As the night wore on, Legg probably sensed he was falling behind, and pictured himself reporting to his midnight relief that he had only accomplished a small part of the list, leaving most of the work for the oncoming cadre, a failure for which he would be held personally responsible. Not surprisingly, Legg's log of the hectic night's work was sparse, as he rushed about the plant trying to complete his work. In the five hours that he stood the watch, Legg made a single written log entry, where he succinctly reported the completion of the first two items on the night orders: "Pumped reactor water to contaminated water tank until reactor water level recorder came on scale. Indicates +5 ft. Replacing plugs, thimbles, etc., to all rods." After making that entry, Legg hurried back to the reactor, undoubtedly hopeful that the third item on the night orders would go smoothly: connecting the rod drive motors.

The USS *United States*: The cancellation of the first "supercarrier" in 1946, five days after her keel was laid, set off the Revolt of the Admirals. (United States Navy)

EBR-1 in Idaho produced usable electricity from nuclear fission for the first time, illuminating four light bulbs, on December 20, 1951. (United States Department of Energy)

Hyman G. Rickover in 1922, the year he graduated from the United States Naval Academy. Asked if he faced anti-Semitism in the Navy, he responded that he'd given most of his antagonists "higher priority reasons to hate me." (Herbert Orth/Time & Life Pictures/Getty Images)

Rickover in 1951, near the time of his promotion battle. His tendency to wear civilian clothes was one of many habits that incensed his opponents.
(Hank Walker/Time & Life Pictures/Getty Images)

President Eisenhower at the United Nations, December 8, 1953: his "Atoms for Peace" speech articulated the hope that atomic power might be used for something other than weapons. (Herb Scharfman/Time& Life Pictures/Getty Images)

Lewis Strauss, Chairman of the Atomic Energy Commission. He forecasted in 1954 that atomic energy would make electricity "too cheap to meter." (Time & Life Pictures/Getty Images)

The christening of the USS *Nautilus,* the world's first nuclear submarine, January 21, 1954, "Something new under the sun." (Keystone/Hulton Archive/ Getty Images)

Colonel James Lampert in 1955, the year ground was broken at SM-1 at Fort Belvoir, Virginia. (Photo courtesy of Hester Hill Schnipper)

General Donald Keirn in 1951. He was in charge of the Air Force's nuclear airplane project until nearly the end. (National Archives)

SL-1: The smallest reactor at the National Reactor Testing Station in Idaho exploded on January 3, 1961, killing three men. It remains the only fatal reactor accident in American history. (United States Department of Energy)

SM-1, Fort Belvoir, Virginia. The Army built its first nuclear power plant just eighteen miles from the White House. (Bettmann/Corbis)

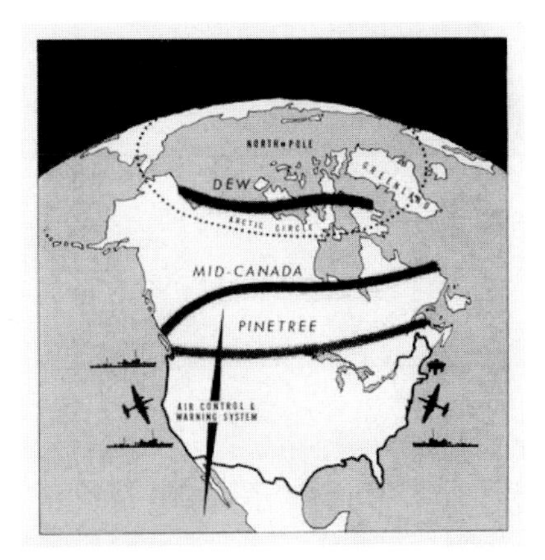

The DEW line was designed to be an impenetrable radar fence through which no Soviet bomber could pass. Many of the remote station were to be nuclear powered. (United States Army Corps of Engineers)

One of the lonely radar stations of the DEW line. The 3,000 mile-long chain of radar stations across the Arctic Circle was one of the biggest construction projects in history. Completed on July 31, 1957, it was obsolete within nine weeks. (Bettmann/Corbis)

The Air Force HTRE experiments joined a nuclear reactor to two General Electric J-47 jet engines. (Todd Tucker)

The NB-36H: The plane flew forty-seven times with an operating nuclear reactor dangling from a single hook in its mid bomb bay. Note the radiation symbol on the tail. (United States Air Force)

The Air Force's $8 million hangar, constructed in 1959, would never house a nuclear-powered airplane. It would store debris from both SL-1 and Three Mile Island. (Todd Tucker)

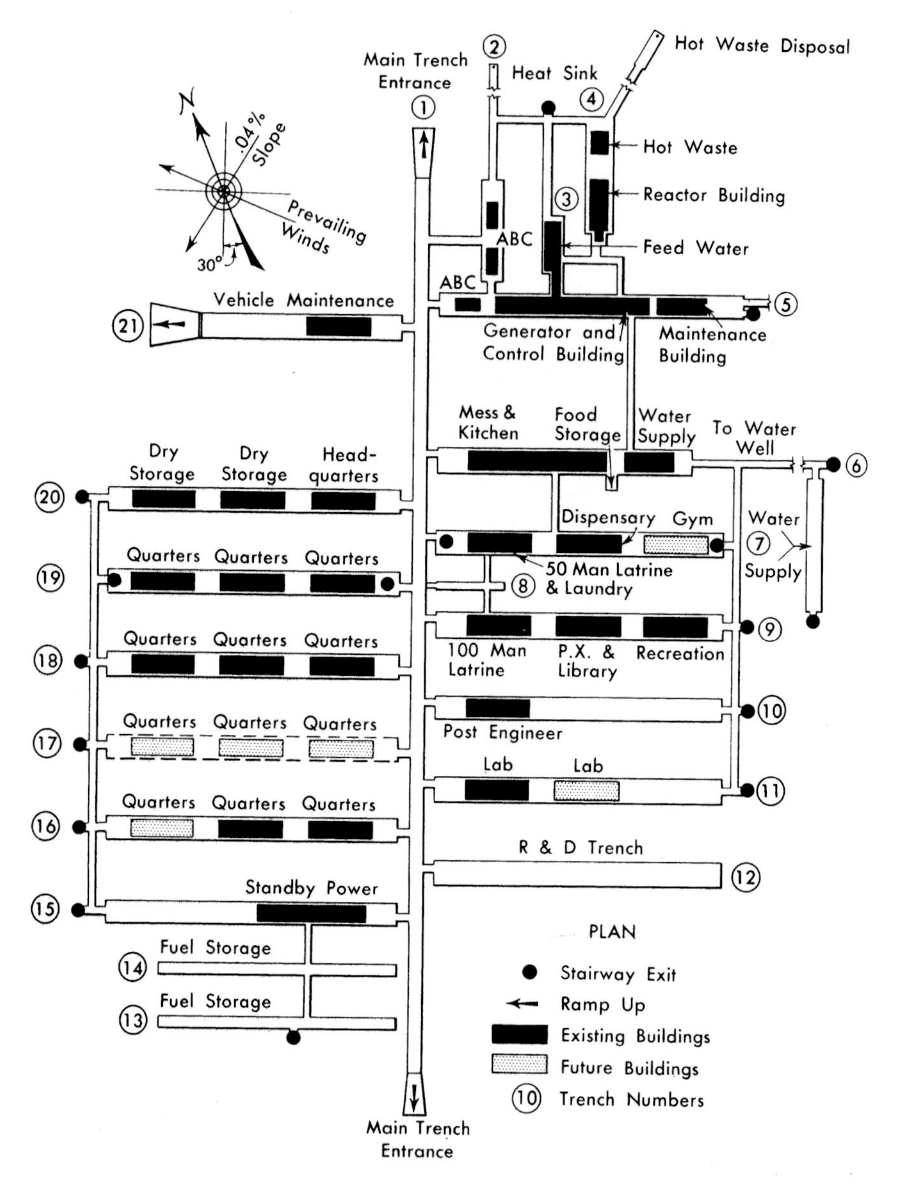

Diagram of the tunnels at Camp Century, Greenland, the Army's nuclear-powered city inside a glacier. Note the reactor at the top of the diagram. The buildings were heated; the tunnels were not. (United States Army)

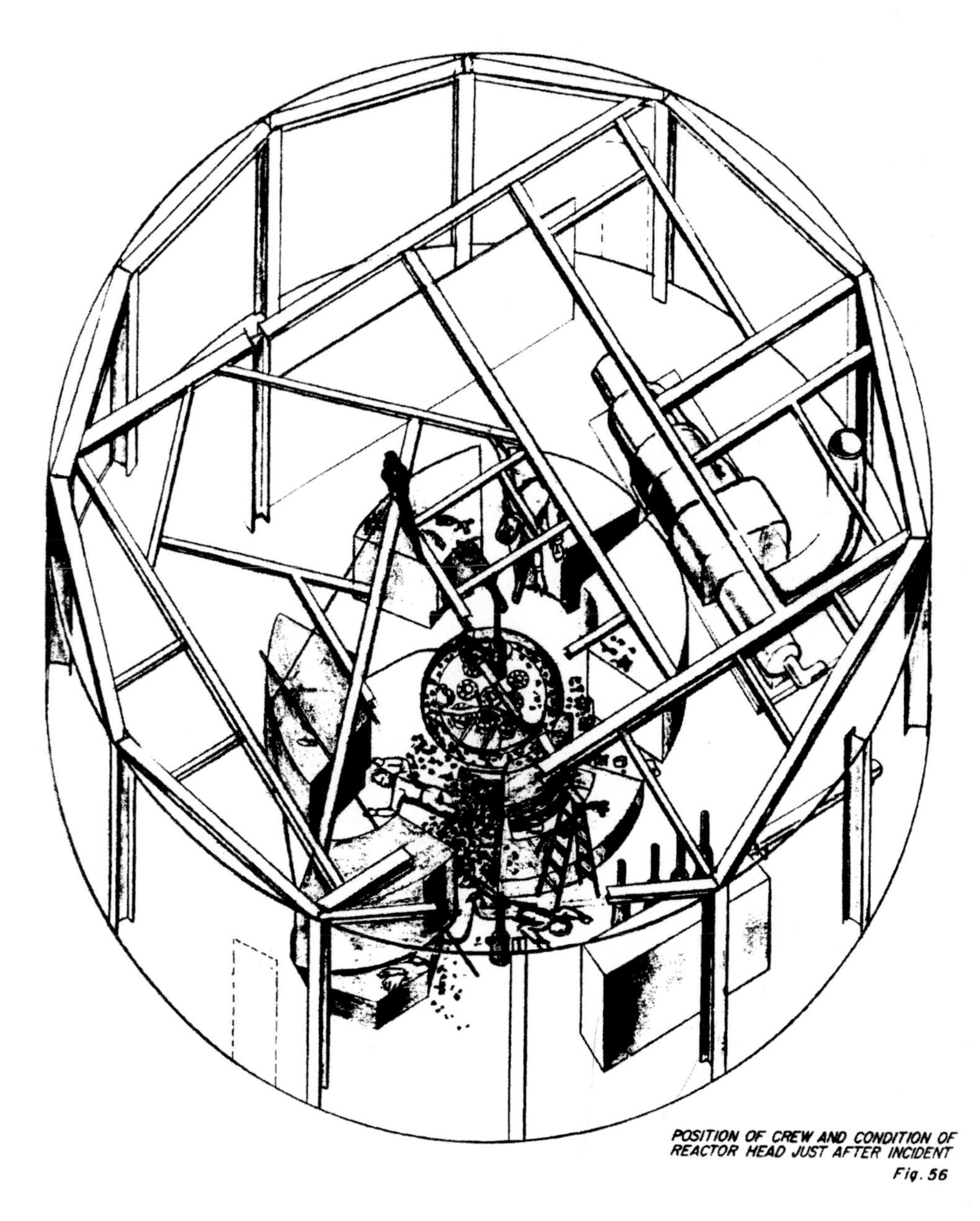

Diagram from government report showing the position of the three dead men. Richard Legg remained pinned to the ceiling for six days while the government designed a safe way to bring him down. (SL-1 Reactor Accident: Interim Report)

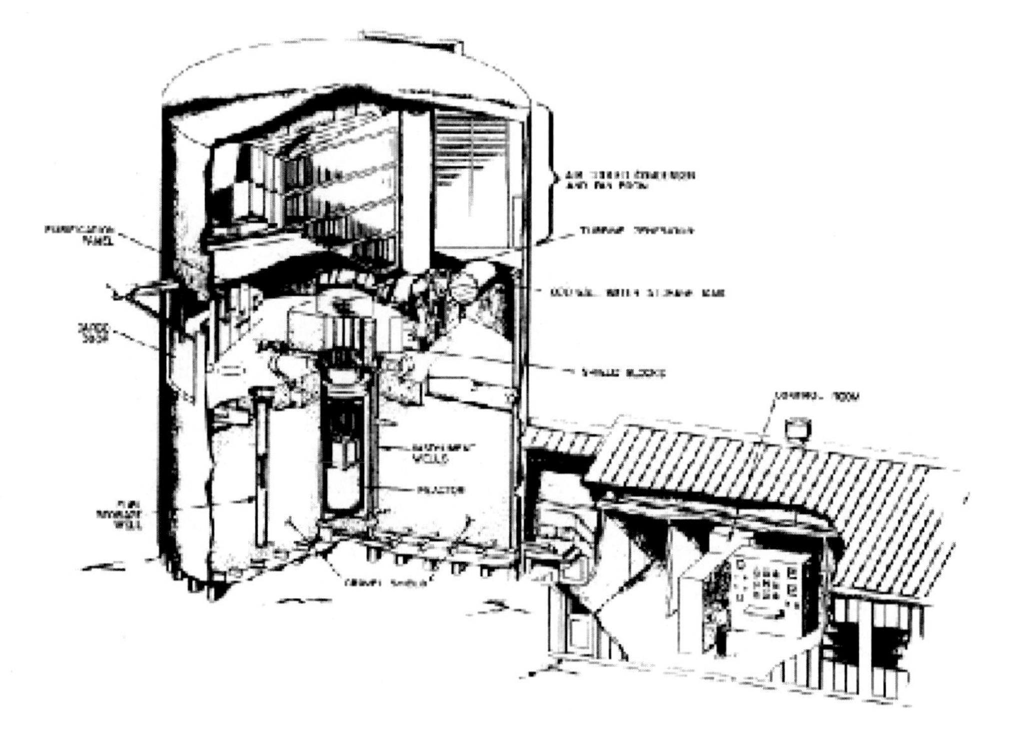

SL-1 (SL-1 Reactor Accident: Interim Report)

The USS *Enterprise,* the USS *Bainbridge,* and the USS *Long Beach.* All three were nuclear powered; the *Enterprise* alone held eight nuclear reactors.
(United States Navy)

The grave of Richard Legg, in Kingston, Michigan, marked by the small American flag. (photo by Louis Wenzlaff)

Rickover's grave in Arlington National Cemetery (Todd Tucker)

THE BACHELOR PARTY

More than a year after the SL-1 incident, Sergeant Paul Conlon, an Army nuke, told an amazing story to Captain R. L. Morgan, the chief of the Idaho Nuclear Power Field Office. Approximately six months before the explosion, Conlon said, he had witnessed the two principal actors in the SL-1 drama, Richard Legg and Jack Byrnes, actually come to blows at a sleazy bachelor party that involved a prostitute. Rumors of all kinds were circulating about the explosion and its victims, but Captain Morgan recognized that this one needed to be investigated. He dutifully alerted his chain of command, and the follow-up was soon assigned to a man who was already an expert on the personal foibles of all the men involved: Leo Miazga.

The bachelor party took place on a Friday: May 27, 1960. It was in honor of the recently engaged W. P. Rauch, a civilian employee of Combustion Engineering, the contractor charged with running SL-1. The party began with more than twenty-five men at the White Elephant in Idaho Falls and then moved to another bar called the Broiler, where the party continued until after midnight. By this time, the revelers had been whittled down to just a hard-drinking few, a group that didn't even include the groom-to-be. They were Sergeant Paul Conlon, Sergeant Gordon Stolla, Roger Young, Jack Byrnes, and Richard Legg.

There was a sixth guest at the party, a woman known only as Mitzi who had joined the group at the Broiler—at the invitation of Byrnes. Mitzi was, in the memorable words of Miazga's report, "a woman of easy virtue." That the married Byrnes would invite Mitzi to join their party was not a surprise to the men who knew him best. He was often seen dancing with women around town, although Roger Young defended his friend's behavior to Miazga by saying that the dancing was innocent. "Army pay," Young told Miazga, "does not permit any high living or extra-marital affairs." At about 1:30 AM, *after pooling their money for a bottle of tequila to take along, the small group moved the party from the Broiler to Young's apartment. Within thirty minutes, the manager of the complex complained about the noise, and the party moved once again, this time to Mitzi's house, on Maple Street in downtown Idaho Falls.*

Mitzi soon offered to have sex with the men at a price of twenty dollars each. The men managed in short order to negotiate that down to two dollars, a price presumably considered reasonable even on Army pay. In the hazy memories of that evening, none of the witnesses could even remember exactly where Mitzi's house was, so they naturally had difficulty remembering exactly who took advantage of Mitzi's offer and who didn't. Sergeant Conlon, for one, told Miazga that "to the best of his knowledge Byrnes did have relations with Mitzi while Legg declined."

This was definitely the dark underbelly of nightlife in staid, Mormon Idaho Falls, but perhaps not all that shocking. Idaho Falls had become a military town, with all the tawdry commerce that accompanies the multimillion-dollar construction contracts lusted after by regional politicians. If the business with Mitzi struck any of the men as shameful, or disturbing, none of them mentioned that in Miazga's report. Miazga himself seemed interested in the transaction only in how it possibly led to an altercation between Byrnes and Legg. Even more importantly, Miazga wanted to assess whether the incident caused any lasting, festering tension.

The altercation happened later that night, presumably after Byrnes and Mitzi completed their business together. Sergeant Conlon was sitting on Mitzi's couch, heard a scuffle behind him, and turned to see that Legg and Byrnes were throwing boozy punches at one another. Conlon attempted to break up the fight by pushing the men outside, destroying Mitzi's screen door in the process. It was enough to stop the fisticuffs. Byrnes stepped back into the house, but Legg left with Conlon, ready to go home to his bride after an eventful night: he had been married to Judy only two months. In the car, Legg didn't mention what his fight with Byrnes was about. Conlon speculated to Miazga that Legg might have made some comment about Byrnes and Mitzi. But he thought it more likely that Byrnes had chastised Legg for drinking more than his share of the communal tequila.

chapter 5

THE AIR FORCE

Hyman Rickover believed that building a nuclear submarine was the ultimate act of willpower, something that had to be wrested into being while struggling against the laws of nature and the idiocy of the military bureaucracy. The Air Force, in contrast, believed that the nuclear-powered airplane was profoundly, beautifully natural, an object so perfect that it would take flight almost spontaneously once all obstacles were removed. The dream of a plane with almost infinite endurance was as seductive to the Air Force as the true submarine had been to the Navy. Air Force Lieutenant General Roscoe Wilson, a Strangelovian character who would appear in most scenes of the nuclear airplane drama, summed up the importance of the atomic plane in 1960: "Our success in weaving the benefits of nuclear propulsion into our present air power concepts and operational forces will in large measure determine the extent to which the United States Air Force will maintain its dominant role in future years." It is safe to assume that General Wilson was referring to the Air Force's dominant role both within the world, and within the American defense budget.

As with the nuclear submarine, the basic theory of the nuclear airplane was deceptively simple: the heat from the combustion of fossil fuel would be replaced with heat generated by nuclear fission. A traditional jet engine works by compressing air into a combustion chamber where it is combined with fuel and ignited. A massive pressure increase results. Some energy is extracted from the gas mixture via a turbine to power the compressor and other systems; the rest is expelled through a rearward-facing nozzle, which creates thrust in the opposite direction. The nuclear airplane the Air Force envisioned would simply replace the heat from burning jet fuel with the heat from a nuclear reactor. If the Navy could do it inside a submarine, then the Air Force was certain they could do the same in an airplane.

The Air Force underestimated both the accomplishments of Admiral Rickover and their own unique technical challenge. Much of that challenge boiled down to the weight and size of a nuclear reactor and its associated shielding. One way of relating the power of any engine to the weight of the vehicle it moves is by calculating the "power loading"—the ratio of a vehicle's weight in pounds divided by the horsepower of its engine. The lower the number, the more relatively powerful its engine. The power loading of the *Nautilus* was around 150, meaning that the nuclear-powered engine room generated one horsepower for every 150 pounds the ship weighed. A supersonic bomber required power loading closer to 4. By another measure, the airplane reactor would need to release around five hundred times more heat energy than the submarine plant. The weight of the power plant combined with the bulky shielding required to keep the crew safe would combine to make an airframe of unprecedented size. Even boosters for the Air Force project predicted that the nuclear-powered airplane would have to weigh at least a half-million pounds. Compare that to the Air Force's B-36 bomber, the giant plane that had precipitated the Revolt of the Admirals—fully loaded with atomic bombs, the B-36 weighed about half that.

And there were obvious safety issues. A catastrophe aboard a nuclear submarine would put the ship on the bottom of the ocean, where it would in effect be shielded by millions of gallons of seawater. As unpleasant and environmentally damaging as that scenario might be, in certain ways there was no better place on the planet to safely store a deranged nuclear system. A malfunctioning nuclear airplane might crash in the middle of Los Angeles or London, or any other city along its flight path, killing thousands and making the area uninhabitable for centuries. Foreign nations might hesitate to allow a nuclear plane on their airfields, or in their airspace.

Almost from the beginning, knowledgeable people saw problems with the entire concept. Robert Oppenheimer and Edward Teller, who agreed on little else, quietly discouraged scientists from participating in the project. Secretary of Defense Charles Wilson memorably phrased the doubts of the Eisenhower administration in a 1953 discussion about the nuclear airplane when he called the yet-to-be-built plane a "shitepoke." The folksy Wilson went on to explain, "that's a great big bird that flies over marshes, you know, that doesn't have much body or speed to it or anything, but it can fly." Much of the battling over the nuclear airplane took the form of the Republican Eisenhower administration fighting to kill the plane while the Air Force allied with the Democratic Congress to keep it alive, as well as the powerful Joint Committee on Atomic Energy, the same group of congressmen who had helped keep Rickover's career alive and were enthusiastic advocates of all things nuclear. Finally, there were the ravenous contractors who soon became addicted to Air Force allocations, and who always promised that a breakthrough in nuclear flight was right around the corner.

The first contract for a study of the nuclear airplane was awarded to Fairchild Engine and Airplane Corporation in 1946. That early program was given the acronym NEPA: Nuclear Energy Propulsion for Aircraft. It was a 1948 study that really launched the serious pursuit of the nuclear airplane, however, and the serious spending of millions of Defense Department dollars. This study came from the same institution that gave birth to the DEW line: the Massachusetts Institute of Technology. There, the "Lexington Group," a gathering of forty top scientists headed by Professor Walter G. Whitman, wrote a report that was emblematic of the Air Force's determination to see the nuclear airplane as inevitable and to interpret every analysis as positive. The Lexington Group outlined the daunting technical problems that needed solving: materials that could bear the enormous temperatures, shielding that would be light enough to fly and still protect the crew, and a reactor compact enough to fit inside an airframe. Despite their many reservations, the study stated that a nuclear-powered airplane might be possible, if the nation was willing to commit to it on a massive scale. They estimated it would take fifteen years and a billion dollars. (The *Nautilus,* by contrast, cost the nation $65 million.) The Air Force seized on the report's lukewarm endorsement. In 1951, the NEPA acronym was officially replaced by ANP: Aircraft Nuclear Propulsion.

The face of the nuclear airplane project was General Donald Keirn. Keirn was a native of Elbert, Colorado, who had attended the Colorado School of Mines for two years. He then gained an appointment to West Point, and placed 100th out of the 300 cadets in the Class of 1929. He was commissioned into the Field Artillery, but gravitated immediately toward flight, receiving his pilot's wings from the Advanced Flying School at Kelly Field, Texas, in 1930. Noting his engineering acumen, the Army sent Keirn to the University of Michigan, where he received a master's degree in aeronautical engineering in 1938. Keirn was detailed to Wright Field, Ohio, where he eventually became chief of the Power Plant Laboratory, where he would spend all of World War II.

At Wright, Keirn was a key player in the development of America's first jet aircraft. While England was still reeling from the Blitz, Keirn made a dramatic, top secret trip to London to observe the early British turbojets developed by Sir Frank Whittle. Keirn escorted a

Whittle engine back to the United States, a trip he made with the engine's plans manacled to his wrist, and personally delivered it all to the General Electric laboratory in West Lynn, Massachusetts. Based on those plans, the United States built its first jet aircraft, the Bell XP-59, which made its maiden flight on October 1, 1942. For his efforts, Keirn won the Most Excellent Order of the British Empire and the Bane Award in 1944 for the most important technical achievement within the Air Technical Service Command.

Like Lampert, Keirn was a West Point graduate, and like Lampert, he would also be a veteran of the Manhattan Project. Keirn became a liaison officer to the Manhattan Engineering District in April 1946, in time to witness the test drops on the Bikini Atoll in that summer. When the Air Force began looking for an ambitious officer with experience in the two new sciences of jet engines and nuclear physics, Keirn was an obvious and well-qualified candidate. He took charge of the nuclear airplane project in 1951, and, like Rickover and Lampert, held positions in both the AEC and the military. Keirn would be in charge of the program almost until its demise.

Like Rickover, Keirn manipulated the press for every advantage, even as he claimed to be uncomfortable with all the attention. Clay Blair described in *The Atomic Submarine and Admiral Rickover* a scene in which the shy Rickover, upon being asked to sit for a photograph for *Life,* shouted into the phone that he didn't want "any damned publicity." These words were written by a *Time* reporter in Rickover's office, typed by Rickover's secretary, and proofread by Rickover's wife. Similarly, Keirn was labeled as a "man of mystery" in a 1955 *New York Times* profile, a man who "turned down all requests for interviews." A 1957 article, also in the *New York Times,* wrote of Keirn, "He is almost unknown. He would like to remain that way." For a man of mystery, Keirn became adept at calling friendly reporters when it suited him. Ominously hinting that the Soviets were close to a nuclear plane of their own soon became a favorite tactic of Keirn and his lieutenants, a tactic that became especially effective after the shock and outrage of Sputnik. (Rickover, for his part, was not above Cold War scare-mongering when it bolstered his cause. As late as 1971, in a speech on the "unmistakable resolve" of the Soviet Union to become "the most powerful maritime force in the world," Rickover said, "the swimming Russian bear is not yet ten feet tall, but he is five feet, eight inches, and growing rapidly.")

While the 1957 *Times* article may have unintentionally highlighted similarities in their methods for dealing with the press, it also detailed some fundamental differences in the two men. Keirn "does not see a parallel between his work and that of Admiral Rickover." He explained to the reporter, "ours is a technically more difficult job." A more leisurely, academic approach than Rickover's was therefore warranted. Said Keirn, "when you're reaching for ideas you can't beat on a table." The article went on, "General Keirn is free from anxiety and strain." Furthermore, "he tries to finish his day's work by 5 P.M.," and never took anything home because it was all "secret." For himself and every one of his subordinates, Rickover viewed the workday quite differently.

The Air Force and General Keirn resolutely refused to believe that the nuclear-powered airplane was impossible, or not worth the trouble. To be against the atomic plane was to be against the Air Force, and to be against the Air Force was to be against America. These tenets of the faith came starkly to light during one of the low points for U.S. Cold War paranoia: the 1954 security clearance hearings for J. Robert Oppenheimer. The gifted physicist, who, more than any other individual, had made the American atomic bomb a reality, stood accused of being a communist sympathizer. While Oppenheimer's opposition to the hydrogen bomb was his principal sin, his open disdain for the nuclear airplane project was also seen as suspicious—part of a "pattern of action that was simply not helpful to the national defense," according to Air Force Lieutenant General Roscoe Wilson.

After first helpfully explaining to the panel that he was a "big bomb man," Wilson testified that Oppenheimer had been unnecessarily skeptical about the prospects for an atomic plane. Worst of all, one senses in reading Wilson's testimony, he had unfairly favored the Navy over the Air Force. "I don't challenge his technical judgment," testified General Wilson about the brilliant scientist, "but at the same time he felt less opposed to the nuclear-powered ships. The Air Force feeling was that at least the same energy should be devoted to both projects." Oppenheimer lost his security clearance, and the nuclear airplane project lived on.

Early on, the Air Force broke down the development work into a kind of competition between two different concepts, much as the Navy had done with Westinghouse's water-cooled versus General Electric's sodium-cooled reactors. For the Air Force, the two competing ideas were the direct cycle versus the indirect cycle engines. In the direct cycle, air would be taken in the jet engine, heated by direct contact with the fuel elements of the reactor, and expelled through the nozzle. In the indirect cycle, heat would be absorbed by some intermediate heat exchanger, which would then heat the air rushing into the jet engine. While more complicated, the indirect cycle would allow the heat from the reactor to be removed by a medium more thermodynamically efficient than air, such as liquid metal. This held out the promise of a smaller reactor, a major benefit. Furthermore, the indirect system could allow one reactor to power several jet engines on a plane. For these reasons, many thoughtful engineers and scientists believed the indirect cycle held the greatest long-term promise for a practical atomic plane. Nonetheless, the simplicity of the direct cycle quickly made it the front runner. The contractor primarily responsible for research into the indirect cycle was Pratt & Whitney; General Electric took leadership over the direct cycle.

Work progressed slowly on the engines, as the engineers and scientists muddled through the incredibly difficult set of challenges the

Air Force had handed them. But the Air Force was soon itching to advance their program with some kind of dramatic demonstration, and at the same time had a real need to evaluate the effects of radiation and shielding on an airborne plane. As a result, the Air Force undertook one of the most hazardous experiments within a very hazardous program: the flight of the NB-36H.

On Labor Day, 1952, a tornado ripped through east Texas, severely damaging several planes at Carswell Air Force Base near Fort Worth. Among those damaged was a B-36, the bane of the admirals and once again the vehicle for the Air Force's loftiest aspirations. The plane's nose had been almost completely destroyed in the storm. The resident manufacturer, Convair, had a bold suggestion for the Air Force. Instead of just repairing the plane, why not convert it into a kind of flying nuclear test bed? The airborne reactors under development were nowhere near ready to propel a real plane, but the Air Force needed to start thinking about the effects of radiation on both the aircrew and other systems of an operating airframe. There was speculation, for example, that intense radiation might degrade the organic compounds used in hydraulic systems. The Air Force agreed to fund Convair's plan.

The damaged B-36, redesignated the NB-36H, was overhauled and equipped with a 1-megawatt, twenty-ton nuclear reactor, which hung from a single hook in the middle bomb bay. The reactor served no function other than to provide radiation. The core was cooled by air that was channeled over it through ducts in the fuselage. The single hook allowed the plane to easily unload the reactor between flights, for storage inside a deep pit in the hangar at Carswell. The hook also would allow the plane to drop the twenty-ton reactor in flight in the event of a catastrophic failure.

The plane was retrofitted with heavy shielding, including a four-ton lead disc between the crew and the reactor. Because of its extreme weight, the shielding only protected the crew, who sat in the front of the plane, essentially in the shadow of the lead disc. Radiation from the reactor streamed unhindered from the sides and back of the aircraft. All crew functions that normally required personnel aft, such as visual inspections of the engines, were replaced with automated systems and television monitors. On September 17, 1955, a nuclear reactor went airborne for the first time.

The plane took off from Carswell and flew directly over Lake Worth, Fort Worth's main water supply. Somewhere over the southwestern desert, the reactor's three control rods were pulled and the reactor was brought to criticality. In case of disaster, a C-97 accompanied the NB-36H and carried a specially trained detachment of Marines. If the reactor were jettisoned, or the plane crashed, the crew was instructed to drop darts with warning signs and smoke bombs, while the intrepid Marines parachuted after it to man a perimeter around the smoking, lethally radioactive reactor. A B-50 also tagged along to measure radiation emitting from the plane. So intense was the radiation field that the crew of the B-50 found they could reliably estimate their distance from the NB-36H by the readings on their radiation gauges. The plane's route took it 365 miles, from its home in Fort Worth to the Air Force base in Roswell, New Mexico.

In all, the NB-36H flew forty-seven times between 1955 and its last flight on March 28, 1957. All of the test pilots were civilians until near the end of the program, when the Air Force thought it wise to give one of their own some experience with the craft. They recruited legendary Air Force test pilot Fitzhugh "Fritz" Fulton, already in Carswell to test the 1,600-mile-per-hour B-58, to be the first military pilot of the NB-36H. He flew it only once. Thankfully, none of the disaster plans were ever necessary, although during one flight one of the smoke bombs did light off inside the plane, causing some panic, but no jettisoned reactor.

Some details of the NB-36H program were declassified early. A 1956 article in the *New York Times* said the test flights would be useful in "designing atomic planes of the future." However, the full story wasn't known to the citizens under the flight path until the *Albuquerque Journal* reported the story in 1987. In one final strange footnote to the NB-36H saga, the plane's hangar at Carswell had been shuttered for many years, unused and off-limits, a designated radiation area. When base workers reopened the hangar in 2005, after a half century, they found inside lush vegetation, full-grown trees, and a family of the largest raccoons any of the workers had ever seen.

The flights of the NB-36H were not the only time the Air Force took radiological risks in order to advance the nuclear airplane. By 1955, General Electric had made small but costly progress in building their direct cycle power plant. GE had devoted its full industrial might to the cause, absorbing millions of government dollars in the process. It had plants working on the project all over the country, most notably at its sprawling jet engine factory in Evendale, Ohio. (Pratt & Whitney's slightly smaller program for the indirect cycle engine was centered around its headquarters in Hartford, Connecticut.) By the time the NB-36H took flight with its twenty-ton reactor dangling in the middle bomb bay, GE was ready to build a rough prototype of its direct cycle engine. Naturally, the experimental reactor would be operated in Idaho.

At the National Reactor Testing Station, the Air Force was eager to set itself apart from its rivals in the Army and Navy. They built their facility at the Test Area North, or TAN, the desolate, northern end of the reservation, about twenty-five miles north of SL-1. The location required the Air Force to build new roads and supply its own, independently staffed fire department, a small concern for the lavishly funded program. The fundamental nature of the experiments also required the Air Force to find itself the most remote corner of a very remote facility.

In addition to the Air Force's desire to separate itself, the isolation was necessary because there was a fundamental difference between the Navy reactors, or any other reactors in Idaho, and the Air Force's reactor that General Electric was assembling. Normal reactors were "closed systems." Fluid circulated through the reactor, absorbed heat and radiation, gave up that heat in some kind of heat exchanger, and then returned to the reactor in a continuous loop. GE's nuclear jet engine, like any jet engine, was a vigorously open system. The cooling medium for the reactor was the atmosphere itself, the "ultimate heat sink" in the parlance of thermodynamics. Air would be scooped into the jet engine, heated up thousands of degrees by the nuclear reactor, and exhausted out the back, all in milliseconds. As a consequence, the highly radioactive fission products that would stay contained in a closed system would pour continuously from the jet engine into the atmosphere, spread like pesticides from a crop duster wherever the nuclear plane flew.

Because of this, the tests in Idaho had to be conducted outdoors, which meant the prototype had to be somewhat mobile, portable enough to move back and forth from the hangar to an outdoor testing station. This was despite the fact that the test assembly GE had constructed weighed 160,000 pounds: it looked like a four-story factory with two jet engines sticking out the back. In the words of D. R. Shoults, GE's general manager for the program, "no attempt was made to restrict the size and weight of the Core Test Facility equipment to approximate a flight version. Rather the assembly was deliberately made large for ease of access and the extra data collection equipment." This was a marked contrast to Rickover's philosophy, which demanded that every piece of the prototype reactor fit inside a submarine-shaped hull.

The hulking Air Force power plant was built to move to and

from the Initial Engine Test Facility, or "IET," on four rows of railroad tracks, towed into position by a heavily shielded locomotive. When in position, the jet engines were ducted to a 150-foot-tall smokestack, so the radioactive gas that roared out of the engine would at least be directed high into the atmosphere. Inevitably, the reactor was assigned an acronym: "HTRE," for "Heat Transfer Reactor Experiment," pronounced "Heater" by all.

The first experiment, "HTRE-1," took the reactor critical on November 4, 1955, and soon after it powered the two modified GE J-47 jet engines that were lashed to it. The hot, radioactive exhaust shot through the smokestacks of the IET and 1,200 feet into the sky, dispersing fission products far and wide enough to be regarded as safe by the standards of the day. The test was an engineering landmark of sorts: it marked the first time a jet engine had ever been powered by an external heat source. HTRE-2, starting in July 1957, involved subjecting an array of different alloys and fuel elements to the 2,800-degree temperature inside the reactor. For HTRE-3, GE constructed an entirely new reactor that was marginally closer to fitting inside an airplane. The power plant was, at least, arranged horizontally. During the course of all the HTRE tests, GE and the Air Force shot shocking amounts of contamination into the Idaho sky: an estimated total of 4.6 million curies. By way of comparison, the "disaster" at Three Mile Island in 1979 emitted roughly half that. But the HTRE tests were no accident. The reactor would produce and emit radioactive contamination as a normal part of its operation.

The designers of the nuclear plane did, however, want to see what would happen in a disaster, if the nuclear plane ever crashed or caught fire. To simulate this catastrophe, the Air Force in the summer of 1958 designed a series of experiments in which they burned to a crisp a series of used nuclear fuel elements in the open air. The official name of the experiment was the "Fission Products Field Release Test," but it was soon informally dubbed "Operation Wiener Roast." In all, the scientists incinerated nine fuel elements, all taken from another reactor at the Idaho site.

A fan-shaped instrument grid two miles long was constructed in the middle of the NRTS. Previous crude experiments attempting to measure the same things had burned highly radioactive, used fuel elements in a section of airplane fuselage filled with kerosene, but the temperatures achieved had not been sufficient to melt the fuel elements or spread any contamination: a reassuring outcome but a failure of the experiment. For Wiener Roast, a special induction furnace was constructed, and in nine separate experiments highly radioactive used fuel elements were incinerated, and a wide, club-shaped swath of contamination was soon mapped in detail across the NRTS. The experiment took place about four miles north of the Central Facilities Area, south of Rickover's Naval Reactors Facility. In at least one of the tests, the plume of contamination shifted unexpectedly and set off radiation alarms at the naval facility, which surely infuriated the admiral.

A secondary purpose of the experiments was to determine how the animals within the test grid might be affected by the release. The animals that populated rural Idaho had long been tested by base scientists. Deer, dairy cows, and even rattlesnakes were occasionally killed and dissected by base scientists to determine if in the course of their grazing and wandering they had somehow been contaminated. The thyroid glands of jackrabbits were especially prized, since the gland accumulated radioactive iodine, just as the human thyroid would. A Dodge station wagon was equipped with a bucket seat on its right fender, on which an intrepid scientist would shoot at rabbits with a 12-gauge shotgun as their colleagues sped them across the lava flows, allowing them to indulge their Wild West fantasies in the name of science.

For Operation Wiener Roast, a more controlled analysis was desired, so random wild rabbits wouldn't do. Scientists wanted to evaluate specifically how airborne radiation from the experiment affected the rabbits, not the contamination they might ingest on plants or lick off their fur. In the first experiment, on July 25, 1958, thirty Dutch rabbits were positioned across the test grid in restraint cages that exposed only their heads and ears. After the experiment, the animals were killed and the radioactivity of their lungs and tracheae measured. Similar measurements were made on dogs and rats during the last seven experiments. The ninth and last incineration occurred on September 26.

The Wiener Roast experiments were in some ways emblematic of the Air Force's efforts throughout the 1950s. They generated huge amounts of radiation, employed hundreds of scientists, engineers, and servicemen, and generated thousands of pages detailing an effort that was, in its own way, dramatic and impressive. The experiments did not, however, in any discernible way, get the world any closer to a nuclear-powered airplane.

The nuclear plane's prospects were not so bleak as to prevent the Navy from attempting to move in on the Air Force's turf. The project's massive budget made it almost inevitable.

The Navy had kicked in a token amount of money for the ANP's budget almost from the beginning: through the life of the ANP program the Navy contributed $14 million, about 2 percent of the total. (In a similar fashion, the Navy maintained a small presence in the Army nuclear power program, the reason Richard Legg was at SL-1.) But starting in 1953, the Navy began to propose in a serious way a nuclear-powered seaplane, a plane that acquired the decidedly unmartial name of *Princess.* The Navy argued that a subsonic, high-endurance plane designed for antisubmarine patrolling was a more useful, and a more practical, atomic plane than the Air Force's prospective high-speed intercontinental bomber. In addition, because a seaplane took off and landed on the ocean, many of the worst safety issues surrounding the atomic plane would be mitigated, as well as the difficulty of obtaining permission to land on foreign airfields. The Air Force recognized with horror that the Navy's argument actually made sense, and the possibility that the Navy might produce the first nuclear airplane was one more scenario that kept the Air Force committed to its own program.

Voices of reason frequently argued that the nuclear airplane program should be focused on research until some of the fundamental engineering problems had been solved. The Air Force, however, remained committed to building something nuclear powered that could actually fly—whether it was militarily useful or not. The battle cry was "fly early," and anyone that advocated anything less ambitious was labeled as a small thinker, a penny-pincher, or worse. Any attempt to refocus their efforts on pure research was seen, at times with some accuracy, as a veiled attempt to kill the program. The Air Force may have been paranoid, but the program did in fact have many real enemies.

From almost the start of his administration, President Eisenhower had quietly sought to kill the ANP program, which he seemed intuitively to recognize was going nowhere. His secretary of defense, Charles Wilson of the "shitepoke" comment, was also a consistent and biting critic. Once, when commenting before Congress about the nature of much of the ANP's research, he said, "I am not interested, as a military project, in why potatoes turn brown when they are fried." Eisenhower could be just as cutting. In 1959, he told his science advisor, Herbert York, that he thought someone might soon propose putting mile-long wings on the ocean liner *Queen Elizabeth* and installing a big enough power plant to make her fly. York begged him not to make the joke in public, for fear that someone might think it a good idea.

Every time Eisenhower came close to killing the program, however, history intervened. The launch of Sputnik in 1957 made it temporarily impossible to take money away from the project, for fear the Soviets would take the lead in another new technology. In 1958 Eisenhower was again on the verge of canceling the program in all but name, denying the Air Force's budget request of $146 million, following the advice of his science advisors who said the "fly early" plan should be shelved until more research had been done on the basic power plant. Such a change in focus would necessarily be accompanied by a large budget cut. General Keirn furiously campaigned to save the program, delivering a speech to the American Ordnance Association in November 1958 in which he said the Air Force program had been the victim of "scientific conservatism," and that they were, as always, "on the threshold of success in various technological areas."

Keirn, like Rickover, had his allies in Congress. Senator Melvin Price, a powerful member of the Joint Committee on Atomic Energy, supported Keirn in his desire not to cut back on the ANP, but instead to spend even more money. Price and his "fly early" allies insisted that the program be focused on actually building a working atomic plane. The senator pointed out that the country had successfully executed similar crash programs in the past, most notably in constructing the *Nautilus*.

Eisenhower pushed on, determined to eliminate this giant line item from the nation's budget once and for all. He chartered his deputy secretary of defense for research and development, Donald Quarles, to conduct an in-depth technical assessment of the program, and Quarles in turn asked the President's Science Advisory Committee to help. Not surprisingly, the committee rapidly determined, in February 1958, that the prospects for an atomic plane were dismal, and doubted that such a plane was even militarily necessary. The committee concurred with the decision to abandon "fly early," scale back the program, and refocus it on developing materials and reactors rather than on the seemingly helpless business of building a working nuclear plane. The administration, it seemed, had finally mustered the wherewithal to eliminate the ANP. But the Air Force was not yet done fighting for the atomic airplane.

On December 1, 1958, *Aviation Week* published a shocking article that declared "a nuclear-powered bomber is being flight tested by the Soviet Union." It was a masterpiece of Cold War scare literature, and the communist plane the article described sounded suspiciously like the one the Air Force desired to build. Two direct cycle nuclear power plants each generated 70,000 pounds of thrust to power the plane that had been "flying in the Moscow area for at least two months." Considering that the entire article was a fabrication, the details were curiously specific: the plane was 195 feet long, and had a seventy-eight-foot wingspan with the trailing edge of the wing angled at fifteen degrees. The nuclear jet's vertical tail rose twenty-two feet above the fuselage, "a typical 'sail' type fin used by Soviet designers to ensure good directional stability."

Soviet political will, too, seemed to be the stuff of the Air Force's dreams, making the atomic plane a program of "top priority and steadfast support" inside the Kremlin. The editors of *Aviation Week*, in an accompanying, outraged editorial, anticipated the response of the Eisenhower administration, "gray voices from high official places attempting to 'pooh pooh' the existence of a Soviet nuclear-powered bomber prototype and coining smooth weasel-worded phrases to deprecate its significance even if its existence is finally admitted, as finally it must be."

General Keirn coyly declined to endorse the specifics of the magazine article, saying "I have only an intuitive feeling myself that they [the Soviets] are quite well along the road." The *Aviation Week* article also quoted Keirn at length explaining the benefits of the nuclear airplane, and by lending his name to it, Keirn tacitly endorsed the rest of the article.

Just as the article predicted, the "gray voices" of the Eisenhower administration did come out, immediately and forcefully, to deny the existence of the Soviet plane. "There is absolutely no intelligence, no reliable evidence of any kind, that indicates the Soviets have flown a nuclear-powered airplane," said President Eisenhower in a press conference days after the article's publication. In the *Aviation Week* article, he had been accused of "technical timidity, penny-pinching, and lack of vision," but in sticking by his guns, Eisenhower was anything but timid. Certain that the Soviet A-plane was nothing but a fantasy of the Air Force's own making, Eisenhower again proposed slashing the ANP budget.

As a last-ditch effort to save the program, Keirn and Price actually flew with Quarles to GE's sprawling plant in Evendale, Ohio, where the GE executives gave a bubbly presentation about the bright, imminent reality of nuclear-powered flight in April 1959. On May 7, the Air Force men closest to the program, Keirn and the ubiquitous General Roscoe Wilson, again met with Quarles back in Washington, arguing still for their "fly early" dreams. The next morning, Quarles's driver came to pick him up at his home—the secretary was to appear on Dave Garroway's *Today* show for an interview. When Quarles failed to appear on the doorstep, the driver entered the house and found him dead in his bed, a victim of a heart attack at the age of sixty-four. Generals Keirn and Wilson immediately recalled that the night before his death, in their last meeting, Quarles had reversed course and began advocating "fly early." There were no other witnesses to this conversion.

Even as the president was trying his best to kill the ANP, the Air Force built what would be the most lasting monument to the nuclear airplane program: a gigantic hangar in Idaho. The landing strip would have to wait, but the Air Force eagerly poured $8 million into a graceful, swooped structure that was one of the few aesthetically pleasing buildings at the NRTS. The hangar, completed in 1959, was built to hold a plane 205 feet long, with a 135-foot wingspan. A heavily shielded, underground tunnel led from the hangar to the control building. It was by far the largest building at NRTS and assured that no one could ever accuse the Air Force of penny-pinching.

While the hangar was undeniably impressive, the Air Force still had nothing resembling an airplane to put inside it, a fact that was getting harder and harder to explain. Since the program had been led almost from the beginning by one man, it became inevitable that he would need to step down. On August 31, 1959, after thirty years in the military and eight in charge of the nuclear airplane, Major General Donald Keirn retired, and was replaced by a protégé, Brigadier General Irving Branch. Keirn insisted that his retirement was "completely voluntary."

By the time of Keirn's retirement, the ANP had swollen into a giant military and industrial project. The program had a dedicated management staff of 175 people, who oversaw the efforts of over seven thousand contractors, the majority of whom worked for GE. The sheer scope of the project was one of the things that made it so difficult to cancel, even as politicians noticed periodically that the government was not getting a great return on its investment in the nuclear airplane. This was especially apparent when comparing the Air Force program to the Navy's. While the Air Force was burning fuel elements in the desert and constructing a giant hangar for a plane that didn't exist, the construction of nuclear ships had become almost routine. By the end of 1960, just five years after the *Nautilus* got "under way on nuclear power," the Navy had fourteen nuclear submarines in operation, twenty-one under construction, and another eleven authorized. In addition, three surface ships were under construction, including the USS *Enterprise,* the first nuclear aircraft carrier, which would by itself contain eight nuclear reactors.

As the critics gained confidence and began to circle, all those many people who depended on the nuclear airplane for their livelihood went into a kind of frenetic overdrive. In a 1958 speech before the Society of Automotive Engineers, J. F. Brady of Convair explained the necessity for the nuclear airplane, and its present, unfortunate state. "I have watched, and had a part in, the hiring of the young brilliant physicist, with the fate of the world resting squarely on his intelligent forehead, only to watch him grow unhappy, disgruntled, and finally quit because of the slow progress, or many program cut backs." Brady went on to blame, of course, the politicians, whose lack of vision was the only thing keeping the atomic plane on the ground.

> *If only our politicians, military leaders, and numerous Department of Defense committees would realize that even with our first crude power plants we can show useful nuclear powered aircraft—and if they would only remember the utterly useless Wright Brothers airplane—and if only they would remember the low performance of our first jet powered aircraft—and if they would realize that these embryo beginnings are developing a knowledge of almost unlimited possibilities—then maybe they would get off their broad backsides and help this country be first with the Nuclear Powered Aircraft.*

With the death of Donald Quarles, Dr. Herbert York became the scientist in charge of the ANP program for the Eisenhower administration. York's scientific credentials were impeccable. He'd been a physicist for the Manhattan Project, and the first director of Lawrence Livermore National Laboratory in California. Like his predecessor, deathbed conversions notwithstanding, he advocated a drastic scale-down and the complete abandonment of "fly early." In his memoir, York summed up his conclusions dryly:

> *While there had been substantial progress in the rate of spending money, there had been precious little progress toward*

> *solving the basic problems which had been recognized in 1948, well over a decade earlier. After all that time and effort, there were still no materials available with which a useful propulsion reactor unit could be built, the problem of crew and cargo shielding had still not been satisfactorily solved, and potential hazards to the public associated with potential accidents of various kinds were still as bad as ever.*

York recommended a small research budget, almost all of which would be absorbed by Pratt & Whitney on the indirect cycle engine, to the tune of approximately $25 million a year. For advocates of the ANP it was a virtual death sentence. By way of comparison, GE alone had been receiving something around $100 million annually in prior years for its work on the plane.

But there was still hope for advocates of the nuclear plane. After all, they had survived budget cutbacks and reorientations in the past. And the Eisenhower administration, after eight years in power, was almost over—so close to the end, in fact, that Eisenhower declined to implement York's recommendations himself, allowing the incoming Kennedy administration to decide the fate of the nuclear airplane. For years, the Democratic Congress had been the program's friend, so with an incoming Democratic president, the Air Force had reason to be hopeful. Some Air Force officers undoubtedly nursed old suspicions, however, since the president-elect was, after all, a Navy man.

Kennedy was inaugurated on January 20, 1961, asking Americans in his address to ask not what their country could do for them, but what they might do for their country. Two months later Kennedy killed the nuclear airplane, not leaving on the table even the paltry $25 million a year York had reluctantly recommended. In his announcement the president succinctly stated his rationale: "the possibility of achieving a militarily useful aircraft in the foreseeable future is still very remote," and "the time has come to reach a clean-cut decision in this matter . . . We propose to terminate development effort on both approaches on the nuclear power plant, comprising reactor and engine, and on the airframe." Adding insult to injury, Kennedy in the same budget doubled the allocation for the Polaris missile—a Navy program.

In 1948, the Lexington Group from MIT had predicted that building a nuclear airplane would require fifteen years and $1 billion. In fact, the program had lasted thirteen years and had cost the nation slightly over that figure. But it never came close to producing a workable airplane.

Back in Idaho, the program was abandoned with dizzying speed. Manuals were left open on their tables at TAN; file cabinets stuffed with paper were abandoned. The ANP had become the largest employer on-site, but there was surprisingly little nervousness among the legions of suddenly idle General Electric employees. Reflecting a different era, GE was proud of having never in its history laid off an engineer. They all trusted that they would soon be detailed to other projects within GE's sprawling industrial kingdom. And, soon enough, GE did receive another large government contract, one that didn't even require their employees to move away from Idaho. On May 5, 1961, General Electric agreed to clean up the disaster site at SL-1.

CAMP CENTURY

In the history of the Army nuclear power program, few men had the breadth of experience of Ed Fedol. Like almost all army nukes, Fedol trained and worked at SM-1 at Fort Belvoir, Virginia. While there, he also qualified on Gas Turbine Test Facility, GTTF, an early nonnuclear Army experiment designed to further development of highly portable gas-cooled reactors. The gas doing the cooling, when the thing was fully developed, would be normal air. Fedol did a brief stint at the doomed SL-1 in Idaho, from October to December 1960, on the same shift as Dick Legg. He attended Richard McKinley's burial at Arlington National Cemetery. Fedol's most challenging tour, however, was at the plant that marked the high-water point of the Army program. It was PM-2A, the nuclear plant that for almost three years powered a secret city inside a glacier in Greenland.

While the dream of the DEW line's impenetrable shield may have died with the launch of Sputnik in 1957, the Army's dream of perfecting an Arctic mission for itself did not. Escalation was a Cold War motif, so if Arctic bases could no longer warn of a nuclear attack, then they would be made able to launch one of their own. With the United States still struggling to perfect a long-range missile that could reach the Soviet Union from its own soil, the Army obligingly suggested that it could lug hundreds of intermediate-range missiles to the polar ice cap, where they would be within range of 80 percent of the enemy's land. The Corps of Engineers would dig tunnels inside the ice, laying railroad tracks on which the missiles could zip from one hidden launch site to another, invisible and invulnerable. The project was called Iceworm, and the power supply for Iceworm would be nuclear fission.

(It is interesting in retrospect how convinced the United States seemed to be, in those early days of the Cold War, that World War III would actually be fought in the cold. The DEW line, Iceworm, and the Army's plans for nuclear-powered snow trains all presupposed that the next war would be fought on ice. The Air Force even developed a special "survival rifle" for its bomber crews that contained a squeeze bar instead of a trigger—designed for a shooter wearing mittens. These plans are especially interesting in light of the fact that much of the actual shooting of the Cold War took place not on the polar ice cap, but in tropical jungles.)

The scale of Project Iceworm was predictably grand. It would deploy six hundred intermediate-range missiles, all moving on railroad tracks beneath the ice cap. The tunnels would be dug deep enough to withstand any Soviet preemptive attack. The network would be controlled by sixty Launch Control Centers. All this would be dispersed over an area of 52,000 square miles, an area roughly the size of North Carolina. Operating and defending this massive installation would be a force of eleven thousand troops, including four hundred Arctic Rangers to defend against overland attacks and two hundred men to operate an air defense missile system. The Army pointed out that Iceworm offered the mobility and dispersion of the submarine-launched Polaris missiles, but didn't require concentrating sixteen missiles in one location, as on a submarine.

To prove that Iceworm could work, the Army first had to prove that it could station hundreds of men for months at a time in the Arctic. The logical starting point was Thule Air Base, a sizable U.S. installation on the northwest coast of Greenland. Before enhancing their presence on the island with nuclear weapons, the U.S. government first had to convince a wary Danish government that it was in their best interest to turn their territory into a potentially key battleground of World War III. One of the more curious arguments the United States gave to the Danes emphasized that a nuclear attack on Greenland would result in more radioactive fallout on the Soviet Union than on Denmark. Eventually, the Danes consented to the U.S. project, reluctantly and secretly.

Fourteen miles east of Thule the Army built Camp Tuto (short for "Thule Take Off"), which marked, literally, the end of the road: wheeled vehicles could travel no farther on the ice sheet. The Army's experimental nuclear-powered camp would be beyond even that, originally planned to be a hundred miles away from Tuto and thus named "Camp Century." In fact, the ideal site eventually chosen for the base was even farther, 138 miles away, just eight hundred miles from the Arctic Circle.

Travel from Tuto to Camp Century required an array of exotic vehicles on tracks and skis, vehicles with names like Beavers, Polecats, Otters, and Weasels. The preferred mode of transportation to and from Camp Century was the "large swing," a giant train of boxcars on skis pulled by a twenty-eight-ton D-8 Caterpillar tractor specially equipped with a massive fuel tank. Crawling across the ice at 3 miles per hour, a large swing could, in perfect weather, make the trip from Tuto to Century in fifty-four hours. The men on the swings stayed in "Wanigans," boxcars equipped with bunks and crude mess facilities. Even in a large swing the journey was hazardous. Fierce storms could scream down unpredictably with seventy-five-mile-per-hour winds. Whiteouts could reduce visibility to nothing. Giant crevasses opened up in the ice without warning and swallowed entire vehicles. Swings stranded by weather or mechanical breakdown could not always call for help. The bizarre atmospheric conditions of the polar regions could wipe out radio communications for weeks at a time. So trying were the long swings that the men on board frequently hallucinated, seeing mirages of churches, houses, and "medium-sized Midwestern cities" rise from the bleak landscape.

Construction began on Camp Century on June 14, 1959. To build it, the Army's Polar Research and Development Center developed a technique soon dubbed "cut and cover." Using a gigantic Swiss snowblower designed to keep Alpine passes clear, the construction crews cut trenches with geometrically straight sides twenty-eight feet deep into the snow. Corrugated metal arches were then placed over the trench, and the blown snow was blown again, back on top of the arches. It rapidly froze, creating a solid roof over the trench. As well as keeping the city hidden, a crucial objective of Iceworm, constructing the base beneath the surface was necessary. Army researchers had learned that no building could survive aboveground, where the constant wind and snow eventually destroyed even the strongest man-made structures.

The main tunnel down the center of Camp Century, "Main Street," was 1,100 feet long. Off the trench ran a number of side tunnels, and at the far northern end was the tunnel that would hold Camp Century's nuclear power plant. Modular buildings were placed in the tunnels to serve as barracks, hospital, recreational facility, and everything else required by a small Army camp of one hundred or so men—twenty-eight buildings in all. Only the interiors of the prefabricated buildings were heated, with weak electric heaters that hung on the walls, and the soldiers were not allowed to maintain their quarters at anything higher than fifty degrees. The tunnels were unheated and stayed much colder, as low as thirty degrees below zero. Everything about life at Camp Century was hard. The sleeping quarters were all on one side of Main Street and all buildings supplied with running water and sewage were on the other, so any trip to the latrine, shower, or mess hall required donning the entire complement of cold-weather gear and an icy walk.

The Army estimated that it would take 1,500 kilowatts of electricity to power Camp Century, or the equivalent of 850,000 gallons of oil per year. The power plant the Army built for Camp Century was a slightly scaled-down version of the successful plant operating at Fort Belvoir. In this case, however, the plant needed to be truly portable, as it would be delivered to its Arctic home one piece at a time. A similar plant for an Army base in Fort Greely, Alaska, had already been started, so the plant at Camp Century would be designated with a "2": PM-2A. Like the plant at Fort Belvoir, it *was built by ALCO, a common denominator in the Army's most successful plants. The total cost to the Army for PM-2A was $3.23 million. This was at the same time the Air Force was paying General Electric $100 million a year to continue development work on the nuclear airplane.*

The plant was shipped from ALCO's plant in Dunkirk, New York, one pallet at a time, via ship, airlift, and heavy swing. The plant mirrored its Virginia cousin in many ways: it was a pressurized, water-cooled, water-moderated reactor. It did depart from SM-1 in Virginia in one significant aspect—since liquid water was scarce at Camp Century and frigid air so plentiful, the plant used air blast coolers to remove heat from its condensers, rather than cool water. The blast coolers were among the biggest components delivered to Greenland.

The plant's first crewmen were all graduates of the training program at Fort Belvoir. They also had gone to the ALCO factory to assist with the trial assembly of the modular plant at that location. Once all the components arrived at Camp Century, it took just seventy-seven days to assemble them, so well designed was it and so well trained the crew. On October 3, 1960, the plant went critical. On November 12, the plant supplied all of the base's power, and the Army had, for the first time, powered a remote base by nuclear energy.

The nuclear plant's air blast coolers were not the only adaptation the Army made to its Arctic environment. Bringing fresh air into the tunnels required ingenuity. In the winter, constant blowing snow made topside vents impractical. In the summer, when outside temperatures could reach the forties, warm air from outside could degrade the integrity of the icy tunnels. To solve this problem, the Army drilled "air wells," holes fourteen inches in diameter and forty feet deep, in each tunnel, each equipped with a fan that sucked fresh, cold air directly from the deep, porous snow. Fresh water was taken directly from the environment in a similar way. A "steam drill" bored a hole four feet in diameter and 165 feet deep directly into the floor of Camp Century. The steam drill had to run constantly to keep the water from refreezing, but the system supplied the men with some of the purest water on earth, up to ten thousand gallons a day. The men were fascinated to learn that the water they were drinking had been frozen since before the Pilgrims landed in the New World.

Ed Fedol reported to Camp Century in October 1961, and immediately fell into the exhausting routine of eight-hour shifts watching over the reactor, huge, joyless meals, and counting the days until his tour was over. Like most of the men, Fedol found that he had to drink large quantities of water to keep from dehydrating in the extremely dry polar air. He was also surprised to discover that despite the severe cold, very few men became ill at Camp Century. The Arctic climate was as tough on viruses and bacteria as it was on human beings.

As Camp Century reached its first birthday, about the same time Fedol reported, the Army discovered that the snow that made up their walls, floor, and ceilings was not fixed, but rather moved a surprising amount, with a force great enough to splinter wooden buildings and twist steel beams. In the cold tunnels, the rate of movement was roughly an inch per month. In warmer areas, however, such as the mess hall and especially the power plant, the rate was much greater. The flow of the snow was of great interest to the resident scientists: they spray painted black grids on the inside of the tunnels, and watched the squares twist and elongate to measure the travel. The creeping snow required constant shaving of the tunnels by the soldiers to keep them in shape, a backbreaking, miserable process in which seventy-five-pound cubes of snow were hacked from the encroaching walls and dragged out of the tunnels on sleds. As much as forty tons of snow were removed from inside Camp Century in this way each week.

Wall shaving not withstanding, diversions for the men of Camp Century were few. There was a four-thousand-volume library, a Spartan recreational room, and movies in the chapel every night. Tours of duty at Camp Century, because of the obvious hardships, were limited to six months, and all who served there were changed by the experience. Vast amounts of food were required not so much by boredom, but by the enormous energy required of men who labored so hard in such cold conditions. A Saturday

Evening Post reporter who visited the base in 1960 commented that it must be the only Army mess hall in the world where GIs were asked, "One steak or two?" The same reporter noted that church attendance was good under the ice, and that many of the soldiers "experience a revival of their spiritual interests out on the lonely glacier."

There were many hardships to working at Camp Century, but operating the reactor was not among them. Ed Fedol remembered the plant as being a "dream to operate," and that the nuclear system was so elegantly designed they could qualify a new operator within two weeks of his arrival. With a technician's keen appreciation, Fedol still fondly remembers PM-2A's elegant valve numbering scheme. The operational statistics of the plant correspond with Fedol's observation. In March and April of 1962, the month Fedol reported, the plant stayed on line 99.7 percent of the time. The plant then began a run of breaking its own records for reliability: 864 straight hours in May 1962, then 1,038 straight hours, then 2,502 hours. On those rare occasions when the plant did have to be shut down, it was usually because of something that had nothing to do with the nuclear system, such as settling tunnels around the reactor that could not be trimmed while the nuclear plant was operating.

Since Project Iceworm was highly classified, the government made much of the "research" role of Camp Century. In one book-length account, a writer enthusiastically communicated the Army's cover story: "there are very few weapons there—only a few rifles for hunting and driving off an infrequent polar bear. Century is not manned by combat troops, or intended for war." The reporters were apparently not shown the test rails in U-shaped tunnels, or told why the Army might be interested in how such vehicles worked beneath the ice. Even the soldiers serving there were unaware of that long-term mission of the base.

The Army's achievement at Camp Century was remarkable, as remarkable in many ways as Rickover's. They had designed a plant that was truly mobile, assembled on site in brutal conditions in just seventy-seven days. The plant then reliably powered a city beneath a glacier for nearly three years. The Army had proven that nuclear power could be a viable substitute for fossil fuels in areas where logistics made that an attractive alternative. Over the course of its life in Greenland, the power provided by PM-2A replaced over one million gallons of fossil fuel.

However easy Camp Century was to operate, Ed Fedol felt no desire to return for another interminable six-month tour on the ice cap. He left after exactly six months in April 1962, and happily reported back to SM-1 in warm Virginia. By August 1963, however, Fedol learned that he was scheduled to rotate back to Greenland for another tour. With a baby on the way and the memories of Camp Century still fresh in his mind, Fedol left the Army instead. His status as a "nuke," however, was to be lifelong, and in the civilian world opportunities were available in warm climates. Just months after leaving the Army, Ed was working in the control room of the southeast United States' first nuclear power plant in Parr, South Carolina, and was on watch in the control room when the plant first went on line on December 18, 1963.

Despite the success of Camp Century, technological progress soon made it obsolete, just as it did the DEW line. Advances in long-range missiles made the investment in Iceworm unnecessary, much to the relief of the Danes, who weren't that keen on the project to begin with. The Army shut down Camp Century during the spring of 1963, and in doing so proved another design requirement of their nuclear plan—the ability to pack it up and move it to another location. The Army notified the Air Force and the Navy that a reliable, portable nuclear plant was available for their use, but no one wanted to adopt it. So, in the summer of 1964, the plant that marked the high point of the Army nuclear program was sent to the only place in the country able or willing to house a highly radioactive reactor without a home: Idaho.

Camp Century was not the Army's only successful field installation, or its longest-running reactor. PM-3A at McMurdo Sound, Antarctica, went critical on March 3, 1962, and powered the research station there for ten years. SM-1A at Fort Greely, Alaska, went critical on March 13, 1962, and also supplied reliable power for a decade. PM-1, on a lonely hill in Sundance, Wyoming, went critical on February 25, 1962, and powered the NORAD radar station there for just over six years. The Army program's final and most powerful plant was located, ironically, on a ship: the "power barge" Sturgis, named for General Samuel Sturgis, who had been chief of engineers during the birth of the Army nuclear program. The 10,000-kilowatt MH-1A aboard the Sturgis supplied electricity to the Canal Zone in Panama until 1976. By the time the MH-1A was decommissioned, it had outlived the Army program, a program that remains best known inside the nuclear power community for the plant that exploded and killed three men in Idaho in 1961.

chapter 6

THE INVESTIGATION

Jack Byrnes pulled the control rod straight up.

Within 380 milliseconds, the core was critical—enough of the all-powerful central rod had been lifted for the neutrons in the core to sustain a chain reaction in the uranium fuel. Byrnes continued to pull the rod up rapidly past this point, pushing the reactor to supercriticality. The massive stored energy inside the nuclear reactor, designed to supply power to a small Army base for years, was released in an instant, most of it in the form of heat, one-half second after Byrnes first began lifting the rod. It was far more heat than the water inside the core could absorb. Power peaked at 19,000 megawatts.

That massive power spike raised the temperature of the fuel to 2,000 degrees Celsius, a temperature at which the central fuel plates simply vaporized. The water closest to the fuel flashed into steam, pushing the water on top of it rapidly upward. This high-pressure water slammed into the lid of the pressure vessel, on which Byrnes and Legg were standing, with a force of 10,000 pounds per square inch. The force against the lid, the "water hammer," was so massive that it actually lifted the entire pressure vessel, a 16.5-foot-high tank that was welded into place. All of the connections were sheered as the 26,000-pound vessel shot nine feet and one inch into the air.

At the same instant, pressure against the vessel lid ejected the control rods and loose shield plugs with a velocity of 85 feet per second. The shield plug for the #7 rod shot straight up, penetrated Richard Legg near his groin, and went completely through his body, exiting through his shoulder. It propelled him straight up, pinning him to the ceiling.

The pressure vessel fell straight back down, landing back inside its support cylinder. Byrnes and Legg were dead. McKinley clung to life but was doomed. The entire episode lasted four seconds.

As in any small town, big news traveled fast in Idaho Falls. Word of the disaster raced across the black lava in a rapidly expanding circle that centered on SL-1. The first to find out were the emergency workers such as Egon Lamprecht who answered the alarms and raced to the scene on the night of January 3. Elsewhere on the site, workers at other plants scratched their heads and wondered why their normally silent radiation alarms would suddenly sound in the middle of that brittle, frigid night. Soon after, in the homes of top site officials and medical personnel, phones rang, and government cars began racing across the desert toward dark offices and frantic checkpoints. Other employees, such as Sharon Peterson, a secretary for Argonne, didn't learn about the accident until Wednesday morning on the way to work, as the government bus she was riding slowed down on Highway 20 at its intersection with Fillmore Avenue, the road that led to the Army reactor. There Sharon saw a hastily established checkpoint surrounded by men in white anticontamination clothing who carried strange instruments. Everyone on the bus contemplated what had happened and traded the latest rumors. Underlying every conversation was real confusion. The scientists and engineers they worked for and revered had always promised that their nuclear reactors were "inherently safe." How could anything have gone so wrong?

One of those true believers racing to the scene was Allan C. Johnson, manager of the entire Idaho site. Although an architect by training, he had found his way into the Manhattan Project during the war, and had been placed in charge of construction at the NRTS in July 1949. Proving himself an able navigator of the many complex technical and political issues that intersected in southeastern Idaho, as well as the strong personalities that populated the place, he was promoted to site manager in April 1954. The disaster at SL-1 would be his biggest test.

The January 3 edition of the Idaho Falls *Post Register* came out before the explosion, but it did foreshadow the incident in a way, with a front page that was a pastiche of Cold War themes: *Laos Intervention . . . Ike Calls for Increase in Readiness . . . Nikita Voices Attack Claim . . . Castro Dares U.S. to Break Off Ties*. The local paper ran its first story about the accident on Wednesday, January 4, and would continue running articles for weeks. The newspaper, and the messages it relayed from Army and AEC officials, were often surprisingly candid, reflecting the general openness of the Army nuclear program. Detailed cutaway diagrams of SL-1 were printed beneath the headline "Where Idaho Reactor Tragedy Happened." Articles accurately informed readers that the three operators killed "were scheduled to continue a job of wiring the control rod." They stated that while the men had died from the effects of the explosion, the radiation was lethal: "Radiation . . . was at such a high level in the reactor building that emergency crews could only enter the building for a minute at a time without exposing themselves to excessive radiation limits."

While frank in most respects, the AEC and lab officials, as well as the local journalists acting on their behalf, were perhaps overly quick to reassure residents that "there is no radiation danger to populated areas of Idaho and Utah." They explained that the explosion at the reactor was "nothing like the explosion of a nuclear bomb." It was, instead, a "nuclear runaway . . . a very sluggish reaction compared to that of a bomb, and neither runaway nor a supercriticality accident could produce a nuclear explosion even remotely approximately that of an atomic bomb."

Apparently not everyone was convinced. Six days after the accident, the *Post Register* attempted again to soothe those locals who suddenly found it worrisome to live next to the largest concentration of nuclear reactors in the world: "Is this fear justifiable? Do reactors blow up like atomic bombs? Can they really spread radioactivity over cities and towns? To all these questions, you can begin with the answer, 'No.'" But the meticulous analysis of the AEC investigators and site personnel proved that radiation and contamination had spread far from SL-1, well outside the boundaries of the testing station, and into the food chain of the area. The SL-1 building, after all, was not a "containment building," meaning it was not designed to withstand pressure from the inside or to keep the results of an explosion contained. It was a deliberate design decision, based on SL-1's portability requirements, its low power, and its proposed remote locations. Given this, the thin metal-walled building actually did a remarkably good job of staying intact and retaining the contamination. Nonetheless, the building was not anywhere near airtight and was equipped with vent fans that exhausted directly to the environment. For days after the explosion, as authorities tried to put together a plan, a stream of highly radioactive contamination from the shattered reactor spewed over eastern Idaho. Unable to contain it, the scientists did what they could to track it.

Sagebrush in the surrounding area was sampled and counted for radiation every day after the incident. The contamination levels on the vegetation rose steadily, until on January 11 the readings were too high to even be counted by the standard method. A map of the sagebrush contamination showed that the radioactive plume from SL-1 had predictably followed the prevailing southward blowing breeze, well outside the boundaries of the NRTS and continuing for hundreds of miles. Thirteen wild jackrabbits were captured and killed for their thyroid glands, all thirteen of which showed levels of radioactive iodine-131 well above normal. Finally, between January 4 and 19, twenty-eight milk samples were taken from five different farms near the southern boundary of the site. Six of the twenty-eight samples showed radioactivity in the milk greater than three standard deviations above background levels. It does not appear any warnings were issued, to area residents or to those farms with the contaminated milk.

The scientists and engineers doing the tests likely sincerely believed what they wrote in their reports: that the radiation levels they detected in sagebrush and milk were nothing to be alarmed about. The levels detected were well below official limits, and, in fact, radiation releases to the environment were an accepted part of life in southeastern Idaho—accepted, at least, by AEC and NRTS officials. Indeed, one of the reasons the analysis of the environment was so efficient in the aftermath of the SL-1 explosion was that radioactive releases from the site's many reactors, whether for maintenance or by accident, were common. The NRTS had an entire system in place for measuring their effects. This reality of Idaho life was dryly described in the IDO report on SL-1: "The reactors and processing plants of the NRTS release under controlled conditions or by accident radioactive gases, liquids, and particulate matter to the air, soil, and water of the NRTS and its environs. These releases result in low-level radioactive contamination of the biota." Many NRTS officials, in short, took comfort in the notion that radioactive releases like the one caused by the explosion of SL-1 were not that uncommon. Area residents, armed with the same knowledge, might have felt differently.

The *Post Register* gave an unintentionally elegant portrait of the nation's nuclear power efforts on January 8, a Sunday. A page one article detailed the latest on the SL-1 disaster, how television cameras would be used to study the damaged Army reactor. In a different article on page 10, David Shaw of General Electric was quoted about the atomic plane less than three months before the program's cancellation: "I want to stress that it is no longer a question of can we build a nuclear powered aircraft propulsion system, but when can we place such a system in an aircraft . . . All nuclear flight systems offer almost unlimited endurance." Finally, in an advertisement near the back of the paper, a Garden City, New York, company offered, for just ten cents, a working model of the USS *George Washington*, "with Polaris missiles that actually fire!" While the Army tried to understand its catastrophe and the Air Force tried to convince the public it was anywhere near a working atomic plane, the *George Washington* was one of fourteen American nuclear submarines actively plying the world's oceans.

The SL-1 disaster made a small splash in the national press. The Associated Press put an article about the accident on the wires; the *New York Times* on January 5 ran it under the headline "3 Killed by Blast in Atom Reactor." A January 13 *Time* magazine article about the incident was headlined "Runaway Reactor." *Time* asked a question that must have alarmed nuclear proponents everywhere: "It was equipped with every built-in safeguard, every 'fail safe' device known to science. What went wrong with SL-1?"

The nation's highest-ranking nuclear officials quickly went to Idaho to try to answer that question for themselves. The *Post Register* reported that AEC General Manager A. R. Luedecke and Commissioner Loren Olson both flew into Idaho Falls on Thursday, two days after the accident. They were joined by Frank Pittman, the AEC's head of reactor development. Olson told reporters at the airport, "we are intensely interested in personally reviewing the incident to learn the facts of the case." The AEC was charged not only with supervising the nation's nuclear reactors, but also with promoting the industry, so in the aftermath of SL-1 its top officials had good reason to be concerned. In many ways, for those promoting the growth of the nuclear power industry, the accident at SL-1 could not have come at a worse time.

On January 3, 1961, despite the years of hype and utopian promises, only three nuclear reactors were actually generating significant amounts of electricity for the commercial grid in the United States. The workhorse of these was Shippingport, Rickover's power plant in Pennsylvania, which had generated the vast majority of the nation's commercial nuclear power since first going critical in 1957. The 110,000-kilowatt Yankee power plant, in Rowe, Massachusetts, was brand-new, having gone critical for the first time in August 1960. The largest commercial nuclear plant so far, at 180,000 kilowatts, was the Dresden, Illinois, plant, located like so many landmarks of the early nuclear age near Chicago. That plant had gone critical for the first time in October 1959. Ominously, the Dresden plant was shut down unexpectedly in November 1960 for what the Atomic Energy Commission called "control rod problems."

And while few in number, those three operating plants were the industry's success stories. In what would become a hallmark of the industry, dozens more were stuck in construction, behind schedule, and vastly over budget. Some, such as a plant in Pasadena, California, were postponed indefinitely because of safety concerns about the location. While the public was still generally accepting of the new technology, there were signs that a grassroots anxiety about nuclear power was building. Boosters were worried that the crisis at SL-1 would portend, or even contribute to, a major reversal in an industry that already seemed to be stalled. *Time* ran an article titled "Atomic Slowdown" in May 1961, which cited the SL-1 accident in its list of problems with the industry.

This anxiety reached its peak during the building of the Fermi plant in Monroe, Michigan, near Detroit. While the plant was under construction, a consortium of labor unions took the government to court, arguing that it was unsafe to locate a nuclear plant near a populated area. The most vocal critic was Walter Reuther, the powerful president of the United Auto Workers. Critics of Reuther said his real complaint was with private development of nuclear power plants—the unions and Reuther advocated complete public ownership, a TVA-style industry. Whatever their real motivations, the unions were able to put together a strong case that included a list of more than forty accidents, major and minor, occurring at nuclear power plants. Shocking many, the U.S. Court of Appeals actually agreed with Reuther, and in June 1960 the court revoked the construction permit for the Fermi plant granted by the AEC. In November, the Supreme Court announced it would review the decision, and the AEC fretted in its annual report at the beginning of 1961 that if the decision were not reversed, "it could be a serious blow to the progress of power reactor development." Everyone involved knew that when the Supreme Court met to review the case, Reuther's updated list of accidents would include the deaths caused and the radiation released by SL-1.

While the physicists and engineers tried to reconstruct what had happened inside SL-1, investigators also needed to look closely at the highly radioactive bodies of the three victims. Few doctors had ever worked in these kinds of conditions. Fortunately, the federal government did have in its charge such a man at Los Alamos, another of the government's desert hiding places for nuclear experimentation. He was Clarence Lushbaugh, a University of Chicago–trained pathologist, and he had the exceedingly rare experience of having performed an autopsy on a radioactive corpse.

That victim was Cecil Kelley, a thirty-eight-year-old father of two, and a civilian employee of the Los Alamos National Laboratory. He was fatally injured in an accident two years before the SL-1 explosion, at 4:35 PM, December 30, 1958, at the very end of the last workday before the start of the New Year holiday. In its timing so near a holiday, the accident mirrored an aspect of SL-1.

Kelley was working with a 225-gallon steel tank, a tank slightly taller than he was, doing the kind of work he'd been doing for most of his eleven years at Los Alamos: recovering trace amounts of valuable, dangerous plutonium from waste materials. The tank held liquid that was supposed to contain an extremely low concentration of plutonium. In fact, the tank contained several different solutions, some of which had concentrations of plutonium about two hundred times higher than was expected or safe. The different liquids had layered themselves in the tank. Kelley, following his procedures, turned on an electric stirrer and watched the contents of the tank through a sight glass, as the liquid swirled into a vortex. That motion was just enough to concentrate the plutonium from the different layers together, a geometry that brought it to criticality for a fraction of a second. A nuclear chain reaction began inside the liquid, the contents of the tank became supercritical, and a flash of blue light, like that of a flashbulb, filled the room.

In that instant, Kelley received a massive radiation dose. He stumbled outside the room, screaming, "I'm burning up!" His fellow workers at first thought he'd been burned by a chemical spill, and rushed him into a shower. When the truth was discovered, he was rushed to the Los Alamos Medical Center, where he quickly spiraled through all the stages of acute radiation poisoning: retching and vomiting, a brief period of coherence when he was able to describe what had happened, and then a complete collapse as almost every function of his body shut down. He died thirty-five hours after his exposure, on the first day of 1959.

Clarence Lushbaugh was the resident pathologist for both the lab and the greater Los Alamos community. Lushbaugh saw in the Kelley tragedy a unique scientific opportunity. Kelley had throughout his career been routinely monitored for plutonium ingestion. Now, his organs could actually be examined in detail, so that scientists could see how precise their estimates had been about the amounts of plutonium absorbed inside his body. In addition, they could see the manner in which plutonium was deposited in Kelley's organs and bones. Without bothering to seek permission or even notify the family, Lushbaugh took from Kelley over eight pounds of organs to analyze, including his brain and spinal cord, which he transported back to his lab in mayonnaise jars.

Lushbaugh became fond of the process. After Kelley's death, the pathologist made it a practice to take tissue samples from every autopsy performed at the Los Alamos Medical Center, even those from people who were not laboratory employees. Just as with Kelley, these studies were performed without the permission of the families involved. The work continued until 1980, affecting an estimated four hundred cadavers. In 1996, a class-action lawsuit was filed on behalf of those families, targeting the University of California, which ran the lab; the medical center; and Dr. Clarence Lushbaugh himself. The lawsuit was initiated by Katie Kelley Mareau, daughter of Cecil Kelley. The medical center and the University of California settled the suit for $9.5 million in 2001, but Lushbaugh had died the year before without settling, or even conceding that he had done anything wrong. In a deposition, when asked who had given him the authority to take eight pounds of organs and tissue from Cecil Kelley, he testified, "God gave me permission."

That controversy was far in the future at the time of the SL-1 explosion. In 1961, because of his experience with Kelley, Lushbaugh was one of few doctors in the world who could claim any experience in the field of examining the dead bodies of radiation victims. Officials in Idaho were grateful for his availability.

The SL-1 victims represented a much greater challenge to Lushbaugh than had Cecil Kelley. Kelley had died purely from a radiation dose. His limbs and body were relatively intact. While some elements inside Kelley's body, such as sodium, had become activated in the incident, Kelley's corpse as a whole was not all that radioactive. The SL-1 victims, in contrast, were ripped apart in the explosion, and their bodies were so radioactive that it wasn't safe to be in the same room with them, much less operate on them with anything resembling normal methods. To perform autopsies on the victims of SL-1, Lushbaugh would have to improvise procedures, construct tools, and just as he had with Cecil Kelley, create no small measure of outrage.

Some wondered if a hazardous, time-consuming autopsy of the three victims at SL-1 was even necessary. After all, there was little doubt about what killed them. There was the predictable argument that such an autopsy would be a learning experience, an argument that would be used to justify virtually every expensive, radioactive procedure during SL-1 recovery. The most pressing objective of the autopsy was more pragmatic. The three bodies in their radioactive state could never be returned to the families for burial—even beneath the ground they would emit far too much radiation to ever be safely interred in a normal community cemetery. A. R. Luedecke actually advocated burying Byrnes, McKinley, and Legg in drums on the guarded, gated grounds of the Idaho desert, treated as so much radioactive waste. Unlike Luedecke, NRTS site manager Allan Johnson knew he would eventually have to face the Idaho Falls community with whatever decision they made. He argued that every possible measure should be taken to prepare the bodies for a return to their families, even if that required some kind of special treatment by the pathologist and specially made shielded caskets. Johnson won out, but that meant that somehow the radiation emitted by all three corpses had to be drastically reduced. During the autopsy, Lushbaugh pursued this objective with brutal efficiency.

Lushbaugh and his team left Los Alamos at 2:30 PM on January 8, 1961, in a military DC-3, and arrived less than four hours later in

Idaho. They found the three heavily mutilated bodies waiting for them inside the decontamination room of the Idaho Chemical Processing Plant. The Chem Plant was a facility designed to recover any trace of U-235 left on ostensibly spent fuel rods. It was also used to extract radioactive lanthanum-140, known as RaLa, a material used in the manufacture of nuclear weapons. Byrnes and McKinley had been placed in stainless-steel tanks filled with alcohol and ice, while Legg, by far the most radioactive, was still in the lead cask that had carried him away from the site after his removal from the SL-1 ceiling. The Chem Plant was an improvised choice—no one had ever thought to create a procedure for dealing with highly radioactive corpses, and certainly no morgue or funeral home in Idaho Falls could cope with them.

If the Chem Plant had never been designed as a radioactive mortuary, it did turn out to be in many ways an ideal location. It was relatively close to SL-1. Its decontamination room, where the autopsies took place, was lined with stainless steel and contained drains and large tanks. A large garage door gave access. An overhead crane traversed the room, which proved extremely useful in moving the bodies from tank to autopsy table and back while maintaining a safe distance. The entire facility had been designed to handle tremendously radioactive materials, which the bodies of Byrnes, Legg, and McKinley now were.

One of the first things Lushbaugh discovered was that the bodies had been mutilated so severely in the explosion that they had been misidentified. The bumblebee tattoo on Legg was one of the factors that made him realize the mistake. With certainty, Lushbaugh now identified the first body removed as Richard McKinley. The second body was John Byrnes, and the third body, the one impaled for five days above the reactor, was Richard Legg. Lushbaugh's description of McKinley's mutilated face makes it clear how the bodies could have been misidentified:

> *The head, which was covered by short brown hair, had a semi-circular sharp wound over the vortex which had penetrated the complete thickness of the scalp. The right lower quadrant of the face had been partially destroyed by a penetrating and avulsing wound, which caused a destructive fracture of the right maxilla, the inferior edge of the right orbit, and fracture of the right mandibular joint. Both eyeballs were flattened and contained no fluid.*

Incredibly, McKinley was the least injured of the three, the man who despite all those wounds actually lived for two hours after the explosion.

Lushbaugh improvised an autopsy table by placing a six-foot-long stainless-steel tray on sawhorses. The bodies could be moved by the crane from the tanks to the table. Even the least radioactive body was far too hazardous to approach from anywhere near the normal position of a medical examiner. Lushbaugh created some crude autopsy tools by welding disposable knives and hooks onto four-foot lengths of galvanized steel pipe.

The actual autopsies of the men, once everything was in position, only took Lushbaugh about fifteen to twenty minutes per man. Each member of Lushbaugh's team wore protective gear during the procedure, including a forty-five-pound lead apron, and a portable lead shield held between them and the body. Speed was necessary because of the extreme radiation levels around the bodies, but it did not prevent Lushbaugh from conducting a thorough examination. Every aspect of the victims, from the texture of their bone marrow to the size of their adrenal glands, was scrutinized in a controlled rush.

The cause of death for each man, as determined by Lushbaugh, reflected the power that had been unleashed at SL-1. Interestingly, each man died from a slightly different cause. McKinley died from the hemorrhages on his left hand and the right side of his face. Byrnes died from striking a flat surface "that fractured his chest and drove a rib through his heart." Legg died instantly "from the destruction of his viscera by the rapidly expanding gases that penetrated his abdominal cavity along with a heavy missile."

Lushbaugh also tried to reconstruct the positions of the men at the instant of the accident, and in doing so, he contradicted almost every version of the incident that would follow. Based on his analysis of the injuries, Lushbaugh placed Legg with his hands on rod 9, with Byrnes standing by watching. Almost every other investigator would conclude that Byrnes had his hands on the rod at the crucial moment, and that whatever happened, whether it was an accident, suicide, or murder, happened at his hands.

No matter who was doing the actual lifting of the rod, Legg and Byrnes would have been extremely close to each other while working on top of the small reactor. One man would have been lifting up the rod while the other removed a C-clamp from it, according to the procedure. Almost all future versions of the story, however, had the volatile Jack Byrnes with his hands on the rod while Legg looked on. Lushbaugh's autopsy contradicted that theory. Both of Byrnes's hands, Lushbaugh detailed in his report, were unmarked: "His uninjured hands must have been up and out of the vortex of the blast or protected by his body." Lushbaugh even staged three men around a mock-up of SL-1 and photographed them to show exactly where he thought the men were positioned, based on their injuries. In those photographs, the Byrnes stand-in does not have his hands on the rod or a C-clamp. His hands are at his side, and he is watching the Legg stand-in, who is hunched over with his hands on rod 9. Even Lushbaugh admitted that his version of events should not be taken as gospel. He wrote in the autopsy report that "this reconstruction scene probably is not exactly correct." Still, it shows the power of the mythology of SL-1 that so few SL-1 storytellers have ever bothered to incorporate Lushbaugh's dramatic staging of the moment before the explosion. The image of Byrnes standing passively with his hands at his sides is not as satisfying as the story of a wild-eyed, heartsick, unstable soldier yanking the rod up, killing himself and his crewmates.

Lushbaugh labored to reduce the radiation of the bodies. A traditional means of decontamination, simply washing and rinsing the bodies in a variety of liquids, detergents, and even citric acid, proved almost completely ineffective. The bodies were raised from the tanks, lowered, and washed again and again in the improvised autopsy room, but the radiation remained dangerously, stubbornly high. This was especially true for Legg's body, which was between a hundred and a thousand times more radioactive than Byrnes or McKinley, depending on which part of the body was measured. Prior to the autopsy, his head was the most radioactive area, giving off a blistering 1,500 R/hour on contact. Even a thick lead casket in a deep grave could not safely contain that kind of radiation. In the end, Lushbaugh found the only way to reduce the radioactivity of the bodies was to remove the most radioactive parts. Large chunks of all three men were sliced, sawed, and hacked off by Lushbaugh, and then placed in a drum and buried in the Idaho desert as radioactive waste. This included McKinley's left hand and Legg's head, severed by Lushbaugh with a 1.5-inch hacksaw blade welded to a ten-foot-long pipe: "a rapid, sharp dissection" in the words of the autopsy report.

That report was classified for years, and even now after a Freedom of Information Act request it is heavily redacted, especially in what seem to be the most gruesome sections. Even so, as with all things about the SL-1 explosion, rumors started circulating almost immediately about Lushbaugh's grisly autopsy, and some of those rumors were at least partly correct. One of the most indignant responses to Lushbaugh's methods came from Atomic Workers Local 2-652, a union that represented about five hundred workers at the Idaho site, almost all of them employees of Phillips Petroleum, a major contractor at the NRTS. All three victims at SL-1 were active-duty military, so none were represented by the union. George Dresich, however, president of the local, took the opportunity to write a report that argued that the SL-1 accident exposed endemic hazards in the atomic industry. While he pointed out in his memo the shaky history of SL-1, the lack of medical facilities for contaminated victims, and even the need for adequate life insurance for atomic workers, he seemed to take most personally the treatment of the three corpses. "They were put in stainless steel sinks in shielded areas, packed in ice, to await disposition. Here they lay without proper burial for a week or two weeks, while medical butchers removed glands, organs, blood and what have you, for study purposes . . . heads, arms and what have you were removed and unceremoniously buried in the hot waste dump at the site." Dresich sent the report to Idaho's congressmen and senators, as well as to Abraham Ribicoff, secretary of the Department of Health, Education, and Welfare.

During and long after Lushbaugh's work, an autopsy of a different sort was performed on the SL-1 plant itself. Combustion Engineering, the prime contractor for the plant, conducted the earliest investigation of the explosion, while General Electric was awarded the contract for the final cleanup and investigation. In parallel with both of these, the Atomic Energy Commission itself, through chief investigator Curtis Nelson, would conduct its own inquiry. Like the gospels of the Bible, the three investigations borrowed heavily from one another and all attempted to tell the same story. Also like the gospels, they differ in revealing ways.

Investigating SL-1 was difficult. The reactor was destroyed. The scene of the accident was lethally radioactive. The sixty-second limit for stays inside the building was hardly conducive to methodical investigation. Worst of all, the only three eyewitnesses to the accident were dead. Despite the challenges, the investigators soon identified two important facts about the explosion that approached scientific certainty, and any credible theory of the SL-1 explosion had to account for both. First, the mechanism of the SL-1 explosion was nuclear. Second, the nuclear surge that caused the explosion was itself caused by raising the central control rod.

That the explosion at SL-1 was nuclear was not a foregone conclusion. While radiation alarms sounded as far as a mile away, the same indications could have resulted from a nonnuclear explosion spreading the radioactive material inside SL-1—the same kind of havoc that would be wreaked by a so-called "dirty bomb." As with any high-pressure industrial system, there were a number of ways the SL-1 vessel might have exploded without nuclear criticality. High-pressure steam, for example, could have built up too high or too fast in a core that was, after all, a kind of steam boiler. A chemical explosion might have caused the same thing—there was speculation early on, for example, that the explosion might have been caused by a chemical reaction between aluminum and boiling water.

Finally, some wondered if the explosion at SL-1 might have been deliberate. A small explosive charge placed under rod 9, they theorized, might have shot it out of the core and caused the nuclear excursion. Thoughts of sabotage, while terrifying in their own way, would at least give some succor to those who wanted to believe accidental nuclear explosions were impossible. And these were, after all, the most paranoid days of the Cold War, when many believed Soviet spies lurked around every corner. The suspicion about chemical explosives and sabotage was summed up in a June 1, 1961, personal letter written by Paul Duckworth, Combustion Engineering's civilian supervisor for SL-1 and one of the first men to arrive at the scene of the accident after the firefighters. Duckworth wrote his letter to Dr. C. Wayne Bills, the site's deputy director of health and head of the local investigation into the explosion. As chief contractor for the running of the plant, Duckworth might have had a vested interest in blaming a saboteur, rather than poor management. In his letter, however, Duckworth seems to have been sincerely alarmed about the possibility. "From information currently available," Duckworth wrote, "a very strong possibility exists that the nuclear excursion originated with a chemical explosion of unknown origin that blew out the No. 9 shield plug and rod."

It appears that Duckworth's concerns were unwelcome, or at least not believed, by Bills, as indicated by the fact that Duckworth summarized his suspicions in a June 6, 1961, memo addressed only "to: file," and that he felt the need to have two witnesses sign his statement, one of whom was Roger Young, a Combustion Engineering employee and Jack Byrnes's best friend in Idaho. Duckworth clearly intended the memo to be proof that he had alerted Bills, and hence the entire power structure in Idaho, about his suspicions of sabotage.

Duckworth noted in his memo that the pathologist's estimate of the positions of the men was inconsistent with the theory that had Byrnes pulling up the rod in a fit of pique or suicidal urges—Duckworth was one of few people to point out the discrepancy. He also outlined a number of aspects of the internal damage to the core that were, in his eyes, inconsistent with the theory of the central rod being pulled out manually. In addition, trying to build the case that saboteurs were on the prowl in Idaho, he wrote that twice in February 1961, prior to movie camera entry for core viewing, "wires were cut to the neutron and gamma recorders at the Control Point."

These suspicions eventually made their way to Allan Johnson, manager of the entire site, who in turn wrote to Curtis Nelson, director of inspection for the Atomic Energy Commission in Washington. Johnson, perhaps feeling that any other action might be regarded as covering up, asked Nelson what he thought about involving law enforcement in what at that point was almost purely an engineering and scientific investigation: "We would appreciate your comment with regard to FBI contact on this matter."

Nelson, perhaps more sensitive to the delicacies of public relations from his office in Washington, suggested to Johnson that an explosives expert hired "on a consultant basis" be brought in initially. "Should there be the slightest inkling from his findings that the destruction of SL-1 was deliberate," Nelson wrote, "the FBI would, of course, be brought into the picture."

Johnson took Nelson's advice. The AEC enlisted the services of the prestigious Poulter Laboratories of the Stanford Research Institute, the country's largest private explosives research facility. Thomas Poulter was an intrepid scientist who in addition to his explosives expertise had been second in command during Admiral Richard Byrd's second Antarctic expedition in 1934. Poulter made the trip to Idaho personally. Over the course of three visits, Poulter inspected every aspect of the reactor he could access, paying special attention to control rod 9, looking closely for any sign that the rod had been blown out of the core by a charge of conventional explosives. The telltale signs Poulter searched for were "elemental carbon . . . nitro or nitrate groups," none of which were found in more than trace quantities at SL-1. In addition, Poulter expertly studied the geometry of the destruction, which also showed no sign of conventional explosives, Duckworth's observations notwithstanding. Poulter concluded definitively in a June 1962 letter, "we are therefore convinced from these facts that there was no sabotage involved in this event of the nature which could have been caused by a chemical type explosion." By the time Poulter made his report, there was, in contrast, mounting evidence that the explosion at SL-1 was entirely nuclear.

To prove definitively that an episode of supercriticality had occurred at SL-1 took some nifty nuclear detective work, analysis that took advantage of some of the key truths of nuclear fission. That a uranium atom placed in a field of neutrons will occasionally be struck and split into other elements is at the heart of what makes nuclear power work. Many other elements also undergo transformations when exposed to a neutron flux, even if they don't fission. They become "activated," or radioactive isotopes, assuming forms rarely seen in nature. Activated isotopes turn into new isotopes or elements as they decay away, spontaneously giving up neutrons, protons, and electrons. Because of this decay, they can be assigned a "half-life," a mathematical attribute that declares how much time must pass before half of a certain isotope has disappeared. For example, if you have sixteen ounces of a material that has a half-life of one day, there will be eight ounces of the material left after the passage of twenty-four hours. After two days, just four ounces. Three days: two ounces remain. A shorter half-life indicates quicker decay. Materials with short half-lives are scarce in nature because they decay away so quickly. Nonradioactive materials have half-lives that are infinite.

To prove that an unintended criticality had taken place at SL-1, investigators needed to identify isotopes with half-lives short enough to make them nonexistent in nature, but at the same time long enough for measurable quantities to be present days after the explosion. They needed isotopes that could only be created in a neutron field, an environment that could occur only if SL-1 had gone critical. If someone set a ton of dynamite under SL-1 and blew it sky high, millions of curies of radiation would be spread far and wide across Idaho, but none of the materials thrown about would transform themselves into these telltale isotopes. The presence of these rare, activated elements would be certain evidence that whatever happened at SL-1 happened in the presence of a neutron field, and thus a critical or supercritical reactor. The investigators at SL-1 swung into work, looking for materials that would bear this distinctive signature. Interestingly, each of the three victims made a personal contribution to this part of the investigation.

From Richard McKinley came a Zippo lighter, practically standard equipment for military men of the era. Inside that lighter was a tiny brass screw that held the flint in place. Brass is an alloy made from copper and zinc. Copper as it occurs in nature is more than two-thirds the stable isotope Cu-63: a nucleus of twenty-nine protons and thirty-four neutrons, 29 + 34 = 63. In the presence of a neutron flux, however, some of the copper absorbs one neutron, becoming Cu-64. Cu-64 is rare and short-lived, with a half-life of less than thirteen hours. The tiny screw from McKinley's lighter was cut in half, and measured twice to prove conclusively that a portion of the screw had become Cu-64, an event that could occur only in the neutron field of a critical reactor. The buckle from the watch strap of John Byrnes was also made of brass, and a similar analysis was performed on it, with the same results. Richard Legg made perhaps the most poignant contribution to the analysis. After his body was removed from the ceiling, his gold wedding ring was pulled from his finger. The ring was blisteringly radioactive: 5 R per hour at first. One quarter of the ring was dissolved in acid and analyzed. Gold, as it occurs in nature, is almost entirely of the isotope gold 197, or Au-197, with 79 protons and 118 neutrons. Just like copper, gold could absorb a neutron, becoming Au-198, an isotope almost never seen in nature with a half-life of 2.7 days. The ring placed on Legg's finger by his Mormon bride proved conclusively that Legg had died in the presence of a supercritical reactor.

Other samples were similarly analyzed: copper wire from a telephone, a zipper pull from McKinley's uniform, and even blood from the three victims. The evidence was consistent across the board. SL-1 had gone supercritical, exposing the men and everything in the vicinity to a strong neutron flux. This same nuclear reactor had boiled almost instantly a huge amount of water in the core, which resulted in a sudden, massive pressure increase, which caused the explosion that did most of the damage. It was this pressure that drove out the shield plug for rod 7, which impaled Legg to the ceiling. It is incorrect, but telling nonetheless, how many people were quick to explain that the accident at SL-1 was "not nuclear—it was a steam explosion," an explanation of the SL-1 accident more prevalent now than it was in 1961. Perhaps that is because people telling the story now sometimes act as nuclear power apologists, whereas in 1961 the feelings were so generally positive about nuclear power that no apology was thought necessary, even in the wake of a deadly explosion. The deaths at SL-1 were seen as tragic, but a necessary price to pay for progress. An editorial in the *Post Register* five days after the accident stated, "In the probing of the accident at the Idaho reactor, scientists will undoubtedly find out something valuable." It went on to call the victims "those astronauts of the reactor world."

To say that the deaths at SL-1 were caused by anything other than a nuclear accident is patently wrong. While the men may have died because of blunt force trauma as steam threw them violently against concrete shielding blocks, or in the case of Richard Legg because he was impaled by a metal rod, radiation would have killed the men in seconds had the explosion not killed them in milliseconds. To say that they were not killed by a nuclear accident is like arguing that a homicide victim wasn't killed by a gun, rather he was killed by a bullet.

Armed with the knowledge that the reactor had gone critical, the investigators needed to determine the cause. From the beginning, there was little doubt that the central rod was the culprit, the only rod with the power to start the reactor all by itself. Since Byrnes, Legg, and McKinley were known to be reassembling the rod drives, and since raising that rod was actually part of the procedure, it was a logical conclusion, supported by all the evidence. The procedure for reassembly of the drives required the control rod be raised "not more than four inches." While that is in the procedure, the limit is not given as a warning, and no consequences for violating that step are given. In fact, the sole "caution" in the procedure is a decidedly non-nuclear one, and appears during the thimble removal, when operators are warned, "Caution; this item is very heavy and cumbersome and must be carefully balanced during removal." The entire procedure is notable for its brevity; it takes just about one full typed page for its

PROCEDURES REPRODUCED FROM AEC REPORT OF JUNE 1961

REMOVE ROD DRIVE MECHANISM

1. Secure feedwater valves to isolate rod drive seals from feedwater pump pressure
2. Disconnect inlet and outlet lines to rod drive seal assemblies
3. Remove tie rod studs
4. Remove seal assembly and place on clean blotter paper
5. Remove pinion shaft extension from thimble. Place on clean blotter paper
6. Remove socket head nuts using Allen wrench and soft hammer
7. Lift off thimble. Caution: this item is very heavy and cumbersome and must be carefully balanced during removal
8. Remove two retaining rings and remove pinions and bearings
9. Secure special tool CRT #1 on top of rack and raise rod not more than 4 inches. Secure "C" clamp to rack at the top of spring housing
10. Remove special tool CRT #1 from rack and remove slotted nut and washer
11. Secure special tool CRT #1 from rack and remove slotted nut and washer
12. Remove 8 socket head cap screws and lift off buffer spring housing and pinion support assembly and place on clean blotter paper
13. Secure two 3/8 inch eye bolts into spring housing. Lift off spring housing and place on clean blotter paper
14. Place special tool CRT #2 over rack and extension rod and secure special tool CRT #1 to rack. Connect special tool CRT #2 to hook of overhead crane and take up weight of rack and extension rod. Rotate special tool in counter-clockwise direction; this action disconnects the split housing from the control rod gripper located at the lower end of the extension rod. The special tools and extension rod are then lifted out by the overhead crane as a single unit.

INSTALLATION OF CONTROL ROD DRIVE

1. Assembly of the rod drive mechanism, replacement of concrete shield blocks and installation of moor and clutch assembly are the reverse of disassembly. Replace all flexitallic gaskets insuring that all mating surfaces are wiped clean with alcohol or other comparable cleaning agent. Particular care should be taken when securing rod drive seal cooling lines and fitting. If not properly fitted up considerable leakage will occur and result in a loss of feedwater and pressure.

fourteen steps. And that procedure is actually for the removal of rod drive mechanisms. Reassembly of the rod drive mechanism was simply, operators were instructed, "the reverse of disassembly."

The death and destruction at SL-1 was thus a matter of inches: the number of inches between the presumably safe four inches allowed in the procedure, and the actual distance the rod was raised. Just how far the rod needed to be raised to cause the explosion was a key piece of information, one that sent the slide rules of the AEC buzzing.

The calculation began by determining the exact power of the SL-1 explosion. Armed with this information, the engineers would be able to calculate how far the central rod had been lifted, and how fast, and with that they could, hopefully, create a hypothesis about how exactly the accident occurred. To calculate the explosive power of SL-1, government scientists created a one-fourth scale model of the SL-1 core at the Aberdeen Proving Ground in Maryland, and blew it up with pentolite, a high explosive. These experiments proved that when the real SL-1 exploded, it was with the same approximate force as thirty-two pounds of high explosives.

It was this force that ejected the shield plugs, one of which impaled Legg to the ceiling. Scientists estimated that the shield plugs were ejected from the core at a speed of 85 feet per second. When the scientists were able to look at the rod 7 shield plug they were interested to see that it had apparently rotated 180 degrees in flight—they theorized initially that this was "due to the rotation imparted by the victim." But as they soon found out, the shield plug had been installed backward in the core, one more sign of slipshod maintenance performed by undersupervised crews. The makeup of the crew was an aspect that went uncommented upon in the investigation. It was taken for granted that the Army manned these crews with young enlisted men of limited experience.

With an accurate picture of the explosive forces developed at SL-1, the scientists could take the next step in the investigation. How far and how fast did the central rod need to be lifted in order to cause that explosion?

To calculate the amount the central rod had been pulled, the investigators relied heavily on the detailed operational records of SL-1, the actual heights the rods had been raised to make the reactor critical in the recent past: the "critical height." On January 3, the central control rod had been raised to its critical height and beyond. How far beyond was a crucial question, as it could make clear whether the accident was innocent, the result of a small deviation from the procedure, or the murderous act of a madman. The answer to this seemingly straightforward question—what was the height of the central rod at the time of the explosion?—turned out to be confusing in a way that is completely typical of the SL-1 body of literature. Various reports by different investigating bodies quote heights of 16 inches, 20 inches, and 24 inches. At times, the language of the scientists and engineers seems almost deliberately opaque. Take this footnote from Combustion Engineering's preliminary report on the incident, a note meant to explain why rod travel distances of up to 30 inches were studied during their analysis:

> *Since the rod, in its disconnected position, is about four inches below the zero of the rod position indicator, these withdrawals*

correspond to "indicated positions" plus 4 inches i.e. a withdrawal of 20 inches from the disconnected position represents a withdrawal of 16 from the "indicated zero" position.

The consensus was that the critical rod height for the central rod alone was 16.7 inches. In other words, withdrawn slowly to that height, the central control rod would have made the reactor just critical, perhaps emitting lethal amounts of radiation to Byrnes, McKinley, and Legg, but not causing an explosion. The height necessary to cause the kind of destruction seen at SL-1 was 20 inches. Even with the 4 inches added to make up for the "four inches below the rod position indicator," it was still less than the 30 inches mentioned in some reports, and much less than the dramatic and frequent reference, "almost full withdrawal" or "nearly the entire length of the rod." Twenty inches is approximately the length from a man's knee to the ground. Moreover, the procedure called for Byrnes to lift the rod slightly more than 4 inches. The fatal movement of the rod, the *excess* distance it had to travel, was less than 16 inches: approximately the length from elbow to thumb. In fact, the distance may have been even less, because of factors such as the deteriorating boron strips that were increasing the reactivity of the core. It is a different picture than the one of a rod pulled impossibly, recklessly high by a wild-eyed John Byrnes, a withdrawal so extreme it could never have been accidental.

Investigators also determined how fast the rod needed to travel that distance, to see if it was even humanly possible to pull the eighty-four-pound rod so high, so fast. Before the reactor became critical at the 16.7-inch position, nothing was happening physically inside the core, so speed was not a key factor; Byrnes could have lazily pulled the rod to 16.7 inches with no ill effects, although it is unlikely he could have done so without Legg noticing. However, from the critical position to the lethal, explosive height of 20 inches, heat and pressure rapidly grew inside the core. If Byrnes had pulled slowly, the reactor would have boiled, fuel would have melted, and the reaction would have destroyed itself before the full measure of the damage was achieved. Using the precise formulas of nuclear physics, scientists estimated with certainty that the rod had to have traveled from the 16.7-inch position to the 20-inch position in less than one-tenth of a second.

To see if it was possible to pull the rod up that fast, the investigators created a full-scale mock-up, complete with an eighty-four-pound rod assembly sticking into a mock core filled with water. They then had a number of men stand atop it and pull the rod, as Jack Byrnes might have, expending "maximum effort" in pulling the rod straight up. In every case, the rod traveled the 3.3 inches well within the required time. They tried the rod-pulling experiment with a number of different scenarios: one man pulling, two men pulling at once, a C-clamp in place, the clamp not in place.

Tellingly, one of the scenarios tested for was "Stuck rod, quickly released." This scenario tested one of the theories going around the base, that Byrnes was trying to break the stuck central rod free at the moment before the explosion. Like all the experiments, this scenario had the rod travel the required distance well within the time required to cause the explosion: in this case, fifty-six milliseconds. That the scientists tested for this scenario shows that they were contemplating a commonly held theory: that SL-1's miserable history of stuck rods might have had something to do with the accident. It was entirely believable, and there was much more historic evidence for it than for any love triangle.

While scientists did account for the possibility of a stuck rod in their investigations, they did not account for the most probable cause of the rod sticking: the crumbling boron strips. Since boron is poison to a nuclear reaction, if it was falling uselessly to the bottom of the core, then the core was that much more "reactive," that much closer to criticality. Since the historical rod data used in their calculations were real, empirical data, investigators could claim that in fact all conditions, including the flaking boron, were accounted for. However, this historical data did not, and could not, account for any boron that crumbled to the bottom of the core during or after the last shutdown. And, in fact, the last shutdown on December 23 had been particularly troublesome. This description is from the preliminary report:

> *For the last reactor shutdown, it was required that each control rod be scrammed individually. With the normal cooling flow to the control rod seal housing, two of the five control rods (Nos. 5 and 9) dropped clean. The remaining three rods, which stuck at various elevations, required a power assist from the rod drive motors to go in.*

Whatever was going on inside the SL-1 core was getting worse. And in addition to fouling the control rods, each atom of boron that dropped to the bottom of the core moved the reactor that much closer to criticality, potentially reducing the 16.7 inches that would have been required to pull the central rod to its explosive height. Because the exact effect of those last chunks of crumbling boron was impossible to calculate, it was almost entirely ignored. At the same time, everyone who knew anything about SL-1 recognized it as a serious safety issue. One report stated succinctly: "Although none of these problems are presumed to have had any influence on the incident itself, the progressively decreasing shutdown margin which resulted from the corrosion and other losses of the poison is not in the best interests of safety."

Almost exactly one year after the explosion, the SL-1 disaster claimed another victim: Allan Johnson's career. On January 1, 1962, the AEC's top man in Idaho resigned. While all involved in the investigation seemed determined to pin the blame for the accident on a rogue operator, the sloppy maintenance and poor design of SL-1 had come vividly to light, and it shocked few that someone near the top would have to pay with his job. In Johnson's nearly eight years as manager, the number of site employees had grown from 1,400 to 4,000, and the number of reactors had grown from seven to thirty. Nonetheless, his last year had been dominated by the investigation and cleanup of SL-1. He told the *Post Register* that for "personal reasons," he would retire.

The romantic lives of the SL-1 victims became subject to speculation almost the instant the explosion occurred: the reason why Leo Miazga was detailed to Idaho to investigate John Byrnes. It is a persistent fact of military history that very often the people involved find it easier and more comforting to believe in the sexual indiscretion of their colleagues than in the failures of their machinery. One modern example was the explosion of a sixteen-inch gun aboard the battleship USS *Iowa* in 1989, a disaster that killed forty-seven men. Within days of the incident, naval investigators implied strongly that the accident was the result of a homosexual love affair and a murder-suicide involving Clayton Hartwig, one of the men killed, and Kendall Truitt, a shipmate who was unhurt in the explosion. Only after this rumor had been allowed to firmly take hold, with essentially no basis in evidence, did the Navy recant. Of the actual causes of the explosion, one was purely technical: the dangerous use of a fast-burning propellant. The other two reasons that were identified sound eerily familiar to students of the SL-1 disaster: inadequate crew training, coupled with a new, inexperienced turret captain.

If Miazga was detailed to Idaho to uncover an illicit romance, then his mission failed. One senses in the terse language of his report, however, that he did find something exceedingly strange in his conversations with the other men of SL-1. No one, it seems, even the most experienced men in the program, thought that yanking the central control rod straight up would create the kind of havoc that it did. It's almost as if they believed the industry propaganda that a nuclear reactor could not possibly explode, and one gets the sense that this particular bit of ignorance in Idaho bothered Miazga a great deal. On five separate occasions, while ostensibly discussing the personal lives of Byrnes and Legg, Miazga asked SL-1 crewmen if they knew that raising the central rod would cause the reactor to explode, and on five separate occasions the men said no. Most, in fact, expressed outright shock that the damage had been so extensive. Roger Young, Byrnes's best friend at the site, told Miazga he "was amazed at the damage wrought." Sergeant Herbert Kappel, who actually wrote the mechanical procedures for the plant, admitted that prior to the accident "he believed a rod could be slowly withdrawn as much as 20 inches with no reaction except building up a heavy radiation field." Sergeant Gordon Stolla concurred, adding that "he doubts that any of the INPFO [Idaho Nuclear Power Field Office] group was aware that serious injury could result from the withdrawal of a rod." He added that it was his belief that the worst result of withdrawal of the number 9 rod would be to make the reactor go critical and that there would be no damage to the individual, "except possibly from exposure to radiation."

R. N. Bishop, who had been in charge of the mechanical activities at SL-1, provided Miazga with the most incredible example of this prevailing belief. He told Miazga that in 1958, a Sergeant Robert Honeycutt had actually pulled a rod out of SL-1, to a height of 30 or 36 inches with no ill effects. In a telling detail about the Army nuclear program, Bishop said that Honeycutt stopped withdrawing the rod only when his warrant officer threw a wrench at him, which struck him on the wrist and caused him to release it. The incident doesn't appear to have really happened, at least not at SL-1—another witness told Miazga the incident had occurred at Fort Belvoir. Nonetheless, Bishop was probably not the only man to hear the story, and it contributed to the belief that raising a single rod at SL-1 to any height was not all that dangerous.

Despite this belief, all the men affirmed they had been warned and trained never to withdraw manually any rod more than 4 inches. Just as universally, they said that no one ever told them why that requirement existed, or what the consequences of violating it might be. However, when Miazga asked to be provided with the examination questions provided to trainees about the handling of control rods, he was told that all examinations had been destroyed.

Byrnes, like many of the men at SL-1, had been heard repeating the boast that if he was at a facility like SL-1 that came under enemy attack, he would destroy it by withdrawing the central rod. It appears likely, based on the overwhelming opinion of his crewmates, that Byrnes probably thought doing so would destroy the reactor, but would not result in an explosion, or even any immediate personal harm.

There were two civilian contractors busily conducting investigations and writing reports about the SL-1 explosion: Combustion Engineering through the early phases of the recovery and General Electric for the final phase. Even though neither company built SL-1, both contractors had a vested interest in blaming something other than faulty equipment. Combustion Engineering was the contractor charged with running the plant, and had a direct responsibility for the safety of SL-1. If there had been something unsafe about the machinery of SL-1, Combustion Engineering should have done something about it. General Electric was affected less directly, but nonetheless had a major interest in the overall success of nuclear power, based on its massive investment in the progress of the industry. That's not to say that either company deliberately obscured evidence, but rather that the investigations were conducted by men who believed in the fundamental beneficence of nuclear power. Their reports reflect this. So while there was no direct evidence of malfeasance or sabotage to which the reports could point, both companies hinted strongly that the disaster could only be the fault of a rogue operator—because, according to them, no other logical explanation existed. An example from Combustion Engineering, in their May 15, 1961, preliminary report, is typical:

> *The assembly of the SL-1 control rod drives requires limited lifting (4 to 6 in.) of the control rod to install a nut and washer. The evidence indicates that the crew was at this stage of the assembly operation when the incident occurred. Presumably, the central control rod (No. 9) was lifted too high for some unexplained reason.*

Another example from the same report:

> *The estimated amount of rod withdrawal required to cause the excursion is large, corresponding to nearly the entire length of the rod, and evidence to establish a reason for such a hypothetical withdrawal is lacking.*

General Electric was similarly coy in its June 27, 1962, report:

> *[It appears] that the operation being carried out at the instant of the accident was to raise the central control rod manually a sufficient amount to remove the C-Clamp. No attempt has been made to suggest possible motivations for pulling the rod 20 inches instead of 2.*

Note the telling exaggeration at the end of General Electric's statement. The height required by the procedure was 4 inches, not 2. It's an example of a tendency many of the reports share, this impulse to maximize the error necessary, to make an accident appear impossible.

Curtis Nelson, the chief investigator of the Atomic Energy Commission, and the recipient of the meticulous reports written by Leo Miazga, was less willing to pawn off responsibility for the accident on any of the three victims. While the AEC in many ways had a large vested interest in the success of nuclear power, just as General Electric and Combustion Engineering did, its top investigators pointed out what few others were saying: that SL-1 was poorly designed, poorly maintained, and had been a disaster waiting to happen. Nelson stated it succinctly in the cover letter of his final report, which he sent to A. R. Luedecke. It was included in the AEC's final report on SL-1, published in June 1961.

> *Although we cannot assign the cause or the responsibility for the explosion to any known or unknown act or condition preceding the incident, it is the judgment of the Board that, before the incident occurred, the condition of the reactor core and the reactor control system had deteriorated to such an extent that a prudent operator would not have allowed operation of the reactor to continue without a thorough analysis and review, and subsequent appropriate corrective action, with respect to the possible consequences or hazards resulting from the known deficiencies.*

Nelson had heard all the love triangle rumors—there is little doubt that he detailed Leo Miazga to Idaho specifically to investigate them. Had they found anything, a single scrap of evidence to point toward an illicit romance, Nelson would have undoubtedly published that fact for his grateful sponsors, who were thirsty for an explanation of the disaster that would not impugn their program or the safety of nuclear power in general. Miazga found out many interesting, disturbing things about Legg, Byrnes, and the SL-1 reactor, but he discovered not one iota of evidence indicating angry lovers or murder-suicide.

The love triangle rumor was natural enough in a tight-knit Idaho Falls community that knew well the personal failings of both Jack Byrnes and Richard Legg. It was also a community of people who needed to believe that the nuclear reactors that dotted their landscape were safe. It was satisfying to believe that the explosion at SL-1 was not the result of poor design or dangerous conditions, but instead the work of one deranged individual. If two of the dead men were connected to each other by sexual impropriety, then so much the better, as impugning the crewmen left intact the reputation of the reactors they left behind.

The rumor usually vaguely implied that one of the men was having an affair with the other's wife, without specifying who was cheating on whom. On close examination, neither adulterous pairing seems to make much sense. Judy Legg at the time of the incident was eight months pregnant. Her husband, Richard Legg, for all his professional and interpersonal problems, had never been accused of adultery. Jack Byrnes was less innocent, perhaps, known to dance with strange women and accused of at least one physical indiscretion, with the woman of "easy virtue" at the bachelor party in 1960. It is implausible, however, that the staid, Mormon Judy Legg would have had anything to do with Byrnes, much less when she was very pregnant.

That is not to say that the troubled Jack Byrnes wasn't capable of pulling the central control rod too far for reasons other than a love affair gone wrong: a fit of anger, to besmirch the reputation of his new supervisor, Legg, or even in a suicide attempt out of despair over his disintegrating marriage. That he might not have been aware that his actions would cause an explosion makes this even more plausible. That leads to a more pressing question than what had been the exact thoughts of John Byrnes on January 3, 1961: Why would a troubled, inexperienced young man like Byrnes even be in position to wreak that kind of havoc? Even if no boron had ever crumbled inside SL-1, Argonne Labs and the Army had constructed a core that could go critical with the motion of a single control rod. They then wrote a procedure that had a man stand atop the core and pull that rod out manually, with the difference between the procedural requirement and criticality being something less than 16 inches. The entire complement charged with this dangerous maintenance consisted of three young enlisted men whose sum total of experience could be measured in months. Add to this the deteriorating condition of the core, which made its overall stability unknown. While there was not a scrap of evidence that a love triangle in any form existed, to look for adultery or even the exact speed and distance the central rod traveled may be missing that point. It is like searching for the exact cigarette butt that starts the forest fire after a yearlong drought. Sooner or later, something was going to start the fire.

THE BURIAL

A Navy plane carried what was left of Richard Legg's body to its final resting place. In a steel casket lined with lead, the body traveled from Pocatello, Idaho, to Saginaw, Michigan, in an R5D cargo plane, under the watchful eye of AEC officials who were tasked with tending the unusual and highly radioactive cargo until it was verified safely in the ground. The local undertaker had to borrow a lift truck powerful enough to carry the heavy load.

Legg's burial took place on January 23, in Kingston, Michigan, the tiny, rural town where the Legg family plot was located. The burial was big news in Kingston, where the stories of nuclear disaster and the presence of dark-suited government officials heightened the drama. The AEC had imposed strict regulations on the gravesite; after all, to bury anything else as radioactive as Legg's body would have required warning signs, fences, and guards. Remarkably, it was left to Louis Legg, Richard's mourning father, to relay the AEC's requirements to the keepers of the local cemetery. The secretary of the Kingston Cemetery Association in turn handwrote a letter to A. R. Luedecke: "I am authorized to assure you that the grave will not be opened at anytime without prior approval of the Commission. I can also assure you that appropriate arrangements have been made under which the Cemetery Association will maintain the grave in perpetuity."

Legg's burial took place in front of a small group of mourners clustered together at the center of the rural cemetery on the eastern edge of town. At the family's request, the steel coffin was lifted from its lead vault for a moment before burial; the nervous AEC officials noted that radiation levels doubled when the coffin was exposed. The service lasted only five minutes. The coffin was lowered into the grave, and then surrounded, per the AEC's requirements, with three feet of concrete. A myth would grow up in later years that the concrete poured on top of the coffin was improperly mixed, causing the coffin to float to the surface in yet another macabre scene. While the extreme cold did make mixing the concrete difficult, no such mistake could have lifted the massively heavy casket.

The circumstances of Legg's death entered Kingston lore, and were on at least one occasion woven into a small town controversy. A developer wanted to create a nineteen-acre lake near the cemetery in 1996. His opponents wrote Michigan's Department of Environmental Quality to protest that the lake might flood the cemetery, which held "a burial containing radioactive material." Nearly a half century later, there's no evidence that Legg's grave poses a hazard to anyone in Kingston. Observant locals, however, point out that the tall evergreens surrounding the Legg plot are peculiarly shaped, stunted and gnarled at the bottom in a way that the cemetery's other trees are not.

The bodies of Richard McKinley and John Byrnes shared a plane together, an Air Force C-54 that took their heavily shielded caskets from the Idaho Falls airport to Griffiss Air Force Base in Rome, New York. There it delivered the body of John Byrnes on January 22, 1961. The plane then flew on to Bolling Air Force Base near Washington, D.C., with McKinley's body. Byrnes was buried in Utica, New York, on January 25. McKinley was buried the same day in Arlington National Cemetery. As with Legg, both burials were observed closely by AEC officials discreetly measuring background radiation with their instruments, keeping the families a safe distance from the coffins, and ensuring that the services were kept to a minimum length. McKinley's service lasted just eight minutes.

There was a fourth burial required, that of the SL-1 plant itself. "Phase One" of the plant's recovery ended with the recovery of Legg's body on

January 9, 1961. After that, a lengthy, deliberative study of the plant was made with remotely operated television and film cameras poking in the building on the ends of booms and dangling from cranes. This "Phase 2" would last until April 1961, during which time no one entered the SL-1 building.

The major goal of Phase 2 was to ensure that the reactor couldn't go critical again. There was a real fear that loose fuel might fall back into the core, reach criticality, and start another chain reaction. Another frightening scenario had the core filling with water, which could, theoretically, in the presence of enough fuel, moderate another chain reaction. A thousand-gallon tank of water in the reactor building was drained, lest that water somehow pour into the core and begin a new disaster. There was at this point so much uncertainty about what had happened at SL-1 that every precaution had to be taken.

When it was determined that the core was completely, permanently subcritical, the final phase of the recovery began. The goals of this phase were daunting: determine the cause of the accident, remove the reactor, demolish the reactor building, and decontaminate the surrounding area. Phase 3 officially started on May 23, 1961, the day a contract was signed between General Electric and the federal government. GE was ideally poised, with a large, experienced workforce already in Idaho, all of them without jobs in the wake of the ANP cancellation in March. Five hundred GE employees gratefully accepted jobs in the SL-1 cleanup. Inside SL-1, they each received their quarterly dose limit of 3 rem in a matter of minutes, after which they would go to the back of the line until three months had passed and they could rush back inside for a few more minutes of work. In an AEC training film made of the cleanup effort, many of the workers are still wearing coveralls with "ANP" emblazoned proudly on the back. The ANP hot shop, approximately forty miles from SL-1 at Test Area North, was originally built to house and maintain a nuclear airplane's engines. It would now be used to house any piece of SL-1 that needed to be saved for the investigation.

As for the rest of the debris, the tons of scrap metal, lumber, and contaminated gravel, officials determined it would be better to create a new burial ground 1,600 feet north of SL-1, rather than transport the vast amounts of radioactive waste to the normal NRTS burial site, sixteen miles away. GE workers dug two trenches and a pit, totaling 81,000 cubic feet, filled it with the remains of SL-1, and then covered it with seven feet of dirt. By the standards of the day, this was deemed sufficient and permanent.

The core itself had to be removed to allow the complete decontamination of the site. This was recognized early on as the most potentially challenging part of the demolition job, because of its size and radioactivity. At first it was feared that workmen would need to approach the core to cut through the numerous pipes and beams that held it in place, a potentially hazardous operation because of the radiation levels and the complexity of the work. Soon, however, it was determined that the blast had done that job for them: all of the connections had been severed when the core shot nine feet into the air on January 3. GE was also working against the clock. A goal of December 1 was set to remove the core, before the worst of winter set in and made transporting it even more difficult.

After exhaustive evaluations and practice sessions with elaborate mock-ups, a seven-foot hole was cut in the ceiling of the reactor building, and a crane carefully moved into place. On November 29, 1961, the crane lifted the core from the reactor building and placed it on the back of a flatbed truck. The core's journey began the next day, November 30.

Moving a radioactive, damaged nuclear reactor forty miles was massively complicated. Traffic had to be completely blocked on the entire route from SL-1 to the ANP hot shop: even twenty-five feet away from the reactor, the radiation level was a dangerous 9 R/hour. Twelve sets of overhead wires along the route that hung below the twenty-five-foot height of the load had to be raised, or lowered all the way to the ground so the caravan could roll over them. At the very end of the route, a short new roadway was constructed into the ANP area. The convoy that traveled with the reactor consisted of more than ten vehicles, including security cars, emergency ser-

vices, a camera crew, and radiological surveying teams. All of them drove the forty miles to Test Area North at a cautious 10 miles per hour, starting the transit at 11:30 AM and finishing at 3:30 PM, when the reactor was finally delivered to the ANP hot shop.

With the removal of the reactor, the demolition back at the SL-1 site picked up speed. GE's mission was to dismantle the reactor building and to rehabilitate all of SL-1's other structures, to ready each "for potential beneficial occupancy." Except for the still significant radiation hazard, the work inside SL-1 became almost janitorial. GE described its cleaning equipment as "brushes, dustpans, brooms, square-nose shovels, 5 gallon buckets." It is a fundamental truth that decontamination doesn't actually rid the world of any radiation. The process is one of gathering, concentrating, shielding, and sequestering that which was previously spread far and wide. But once it is created, the only thing that can really reduce the sum total of radiation is the passage of time, the relentless truths of half-lives and decay.

The silo-shaped reactor building was dismantled and the gravel and dirt beneath it methodically dug up and removed to the SL-1 burial ground. The remaining buildings at SL-1 remained in place, and were painstakingly decontaminated, with every individual surface surveyed, cleaned, and surveyed again, until each building was declared completely safe. The accepted practices of the day can seem makeshift in retrospect: contaminated soil that couldn't be carted away was paved over. Floors that couldn't be scraped or steamed clean were covered in a fresh layer of concrete. And contaminated walls that couldn't be scrubbed clean were covered with a thick coat of heavy metallic paint to fix the contamination in place. It would have been far cheaper to raze the buildings and bury their contents, but like so many of the other exorbitantly expensive decontamination jobs resulting from SL-1, like the Pontiac ambulance, it was deemed a worthwhile evolution because of the rare experience the teams involved would gain. GE officially completed the job on July 27, 1962.

The decontaminated buildings of SL-1 stood for decades after the explosion, support buildings for a reactor that no longer existed. The broad area came to be identified as "ARA" for Auxiliary Reactor Area, and the site of SL-1 was ARA II. It lived on the edge of the collective memory in Idaho, a reminder of what could go wrong, a monument to worst-case scenarios. The three main buildings at SL-1 were used as a welding shop and offices until 1984, when they were closed for good, and given a new label, that of "surplus facilities," by the federal bureaucracy.

In 1985, NRTS officials determined that the time had come to completely destroy every building at the site, including their concrete floors and foundations. It is a misconception of many antinuclear constituencies that "industry" monolithically supports nuclear power and therefore would always propagandize to minimize the risks. In fact, there is a robust industry that supports the cleanup of industrial and radiological hazards, and the more menacing the hazards appear, the more extravagant a cleanup is justified. The lingering hazards of SL-1, more than twenty years after the explosion, were deemed once again worthy of a massive, expensive cleanup.

A plan was devised to "decontaminate and dismantle" all that remained of SL-1, a job that would end with covering the area with "clean, compacted soil" to reduce the dispersion of any remaining contamination. The goal, once again, would be to eliminate all traces of SL-1, even those detectable only with the most sensitive Geiger counter. The wheels of the federal bureaucracy turn slowly, however, so it was not until 1993 that the "D&D" job of SL-1 began. Once again the contamination generated in that four seconds in 1961 would create years of work for a government contractor, this time Lockheed Martin. Their engineers documented their efforts in meticulous, voluminous reports. "Through 21 months of field operations," detailed Lockheed Martin in 1995, "38,667 direct man hours have been spent on ARA II D&D activities."

In those first twenty-one months, Lockheed Martin employees chopped up a 50,000-gallon water tank, a 1,400-gallon fuel oil tank, a 1,000-gallon detention tank, and three septic tanks with something called "a mechanical nibbler." While GE had declared the area completely decontaminated in

1962, their successors from Lockheed Martin in 1993 found contamination frequently during the demolition; it had settled behind gypsum wallboard and inside attic spaces, and turned into sludge inside the underground tanks. All the waste was compacted or solidified and shipped to one of the Idaho site's low-level waste storage facilities. The job wasn't completed for four years, when in 1997 the SL-1 site was once again declared decontaminated and safe, every building erased from the Snake River Plain, every spec of contamination presumably sequestered forever.

In 2003, they would try again. A new kind of superdump was invented, a "leak proof landfill," called a CERCLA Disposal Facility, for Comprehensive Environmental Response, Compensation, and Liability Act. (CERCLA is the official name for the act commonly called "Superfund.") The new dump was constructed with two layers of plastic surrounding a layer of clay, and the soil from SL-1 was one of the first things dumped into the 510,000-cubic-yard pit.

The entire site must review its CERCLA compliance every five years, and if any trace of SL-1 peeks out from the edges of its latest home, a new contract will surely be signed, and a new, expensive form of remediation will begin once again.

chapter 7

ENTERPRISE

On September 24, 1960, three months before the explosion at SL-1, Rickover launched his biggest ship: the aircraft carrier USS *Enterprise*. All memories of the USS *United States* were lost in the swirling celebration, even though the *Enterprise* was constructed in Shipway No. 11 at Newport News, the very place where the keel of the *United States* had been laid down five days before her cancellation. Perhaps that's because the *Enterprise* so clearly exceeded even the most grandiose claims of those admirals who staked their careers on the supercarrier back in 1949. The *Enterprise* was longer than the planned length of the *United States* (by thirty-five feet), wider (by seven feet), and displaced ten thousand tons more seawater. And unlike the *United States*, which would have been forever shackled to her support ships and their dirty, flammable, sloshing cargo of oil, the eight nuclear reactors of the *Enterprise* could keep her at sea for years without refueling. Her designers estimated that she would travel 200,000 miles, or eight times around the world, with just the fuel in her initial loading. They were conservative by 7,000 miles. Despite his embrace of the aircraft carrier, Rickover resembled not at all those suntanned, strapping aviators who would have presumptuously named their carrier for the

Republic. The name of Rickover's carrier sounded like a puritan admonition, a summation of his personal philosophy and a charge to all who would board her.

Rickover's ascendance in the Navy, and the seemingly perpetual grip he maintained on it, was not without critics. His powerful advocacy of nuclear propulsion was to the exclusion of all other forms of sea power, and woe to the man who crossed Rickover and suggested that in some cases diesel engines or gas turbines might be worthy investments. Because of Rickover's power, in the Navy and in Congress, and his many admirers, few people ever felt comfortable voicing their concerns out loud. One exception was Elmo Zumwalt, the officer who spurned Rickover's invitation to join the nuclear fraternity in 1959. Zumwalt felt his career was temporarily affected by the decision, but he would go on to become the highest-ranking officer in the Navy, the chief of naval operations, in 1970, the man Admiral Rickover ostensibly reported to. Zumwalt felt, sensibly, that the Navy needed a variety of ships. Zumwalt felt especially strongly that the Navy needed more ships, and that quantity could be increased if not every ship was an expensive nuclear one. His battles with Rickover were epic. In his memoir, he titles his chapter on the subject "The Rickover Complication," and he calls Rickover "a persistent and formidable obstacle to my plans for modernizing the Navy." He also gives an eloquent description of the domain Rickover had constructed for himself in the seventeen years since his promotion fight.

> *I knew that his Division of Nuclear Propulsion was a totalitarian mini-state whose citizens—and that included not just his headquarters staff but anybody who engaged in building, maintaining, or manning nuclear vessels—did what the Leader told them to, Navy Regulations notwithstanding, or suffered condign punishment.*

Admiral Zumwalt would retire from the Navy after a standard four-year tour as CNO in 1974. Rickover continued on, immune to the normal career progression that made every one of his foes merely temporary.

The *Enterprise* carried inside her single gray hull as many working reactors as the Army nuclear power program would build in its entire history, far more than the Air Force would ever construct, and more than the entire U.S. civilian industry would manage to operate for another three years: that's when Ed Fedol, former Army nuke, put the Parr, South Carolina, plant on line on December 18, 1963. At the time of the SL-1 explosion, no one man had a greater personal stake in the success of nuclear power than Hyman Rickover. And while the admiral had no official role with SL-1 or the Army program, he wanted to know why the explosion had happened, to see clearly the real causes. Nuclear accidents were exceedingly rare, and while the admiral undoubtedly had inklings already about the vast differences between his program and the Army's, he couldn't rule out the possibility that he might learn something from the Army's misfortune. And experience had taught him not to wait for the official reports when seeking an unvarnished view of the truth. So while General Electric, Combustion Engineering, and the Atomic Energy Commission sent investigators to Idaho, Rickover commissioned one of his own. He put in charge one of his most trusted civilian protégés, thirty-one-year-old test engineer Clay Condit, who was in Pittsburgh at the time, working at Westinghouse's Bettis Labs. Condit left immediately for Idaho. While Condit's paychecks may have come from Westinghouse, he, like everyone else associated with naval nuclear power, knew who the real boss was.

Acting as Rickover's proxy at the scene, Condit quickly learned the broad outlines of the tragedy. Accustomed as he was to Rickover's iron rule and relentless pursuit of perfection, Condit was shocked by what he discovered. While the document remains classified, like almost everything connected to the naval program, Condit remembers frequently using the word "appalling" in his report. The ongoing history of stuck rods was the most heinous failure, along with the resigned tolerance for the condition. Control rods at SL-1 had stuck sixty-three times; Condit was certain that after no more than three such instances at any of his reactors, Rickover would have ordered not only a reactor shutdown, but also a pound of flesh from all appropriate parties, be they civilian or military. The condition of the tack-welded boron strips was only slightly less galling to Condit, that the Army would allow the reactivity of the core to hinge on such obviously faulty workmanship. Condit, like everyone else in Idaho, heard the rumors of the love triangle and suicide, but to him, the poor condition of SL-1 and it lax supervision was infinitely more scandalous.

Condit dutifully turned the results of his investigation over to Rickover, results that confirmed what many in the Navy program suspected: SL-1 was a sloppy operation, the explosion was a direct result, and such a disaster could never happen in the Navy. A different leader might have gratefully accepted those results and gone on with life, happy to have his methods validated. But Rickover once again surprised many of his subordinates with a course of action so extreme as to seem almost irrational. Rickover called Condit back to S1W, the *Nautilus* prototype at Idaho, the place where Rickover had once felt "real elation." Rickover's mandate to Condit: test every extreme condition, every possible casualty that could ever befall his reactor, tests so arduous that they eventually resulted in the virtual destruction of the plant. While Rickover never stated that the testing at S1W was a direct result of the SL-1 accident, the timing of it was conspicuous, as was the fact that the same man, Clay Condit, was in charge of both the Navy's investigation of SL-1 and the testing at S1W. In addition, the testing included several different reactivity insertions, the exact type of casualty that had destroyed SL-1 and killed three men. The final test sent a slug of cold water into the reactor 250 degrees colder than the operating temperature. The thermal shock caused the thin tubes of the starboard steam generator to burst, shutting the plant down for nearly two years while the steam generator was torn out and replaced.

The other investigators at SL-1 had also drawn conclusions that would have lasting consequences for their respective organizations. Despite the convenient consensus that the accident was caused by an inexplicable withdrawal of the central rod by a rogue operator, the investigators could see, just as Clay Condit had, that SL-1 had been fatally flawed, and that those who tended the reactor had tolerated it.

General Electric published its final conclusions in a November 1962 report, *Additional Analysis of the SL-1 Excursion*, four months after the completion of Phase 3 and nearly two years after the explosion. While careful not to deviate from the central tenet of a rogue operator, the GE investigators could not help but comment on the general problems at SL-1. In the words of the GE engineers, "certain conditions were observed that while not directly associated with the accident, seem worthy of comment."

The foremost condition listed had to do not with the troublesome control rods but with the design of the reactor. A reactor's design, said the report, "should not permit criticality as the result of the movement of a single control mechanism," as was allowed at SL-1. Second, the GE engineers recognized that to allow operators to manually withdraw that rod as part of a procedure, with no safety mechanism preventing full withdrawal, was inviting disaster: "Such movements should not be a part of routine maintenance." Other problems were cited—the disintegrating boron strips, the lack of audible alarms, even the thickness of the fuel cladding. GE summed up its conclusions by referring to the plant's "abnormal conditions."

> *The SL-1 had experienced a history of boron loss, fuel element sticking, and control rod sticking which were of concern to*

> *the operators of the plant and which were undergoing active investigation. Though these problems had no direct relation on the SL-1 accident, it appears in retrospect that they may have warranted plant shutdown, and that continued operation may have indicated unwise emphasis on achieving operational goals.*

The report from the Atomic Energy Commission, authored substantially by Curtis Nelson and informed by the work of Leo Miazga, was far more blunt. As already seen, the report included Nelson's cover letter, which stated that a "prudent operator would not have allowed operation of the reactor to continue without a thorough analysis and review." Nelson's letter also included a scathing critique of Combustion Engineering's performance as managing contractor, as well as this unambiguous declaration, a conclusion that has been ignored in almost every subsequent account of what went wrong at SL-1:

> *The immediate responsibility for the SL-1 incident, still in light of the foregoing discussion, was that of the contractor, in that the contractor was on site and had immediate responsibility for all reactor operations. We specifically absolve the military cadre, as such, from any responsibility.*

Within a few years, however, in popular accounts of the SL-1 explosion the military cadre had assumed complete responsibility for the incident, and in most accounts the love triangle rumor was presented as an accepted fact. These accounts were given credibility when a memo written by AEC bureaucrat Stephen Hanauer was leaked to the press in 1979; the memo repeated the love triangle myth. Hanauer later confirmed to author William McKeown that his memo was just a written account of the rumor that had circulated through the organization for years. As an AEC staffer, however, his memo lent credence to a version of the SL-1 story that was already well on its way to becoming the standard explanation. A half century after the accident, it is almost impossible to find an account of SL-1 that doesn't mention some version of the myth. One example is the 1982 antinuclear book *The Cult of the Atom:*

> *A mentally unstable operator, according to the A.E.C.'s private speculations about the incident, had deliberately withdrawn the reactor's central control rod in order to cause a runaway chain reaction. He was overwrought, officials believed, because he thought that his wife was having an affair with one of his fellow operators.*

This is an especially interesting example, because the book is very much antinuclear: it would seem that a writer with such a point of view would prefer to point out the real lessons of SL-1, the dangers of poor design and the myth of the "inherently safe." Such is the power of the love triangle story that the author chose to highlight it anyway. Other examples are plentiful. From a 1992 article in the *New York Times,* a retrospective on the history of nuclear power:

> *The Stationary Low-Power Reactor, Jan, 3, 1961: The Stationary Low-Power Reactor (SL-1), an Army prototype near Idaho Falls, exploded, killing three technicians and exposing dozens more to radiation. Investigators later theorized that it was a murder-suicide.*

From the *Seattle Times* in 2000:

> *Investigations following the explosion failed to conclude what happened that night, except that a control rod got pulled out*

> *too far and caused an instantaneous steam explosion. Workers had earlier reported the reactor control rod was sticking. But to complicate matters, two of the three men who died that night were in love with the same woman.*

The *Salt Lake Tribune* in 2001 gives a story that mentions the love triangle, only to refute it with another piece of SL-1 misinformation:

> *Rumors circulated that the accident was really a murder-suicide triggered by a "love triangle." After a lengthy investigation, a team of scientists offered their best guess: One of the employees, apparently a Homer Simpson prototype, "goosed" the man who was withdrawing the control rod.*

No scrap of evidence for a love triangle was ever found, but the story has proven durable, perhaps simply because it is easier to understand than stuck control rods or crumbling boron, and perhaps also because so many people close to the incident needed to blame something other than poor engineering and supervision. Whatever the reason, it is hard to find an account of SL-1 that does not include the love triangle story, and nearly impossible to find anything that discusses the decrepit state of SL-1 in January 1961, and the vague procedures that left the reactor quivering on the edge of disaster.

With most parties blaming either a love-crazed soldier or a negligent contractor, the Army nuclear power program continued for a short time, and its successes at places like Camp Century, Greenland, and Fort Greely, Alaska, indicated that perhaps SL-1 was an anomaly in a program that was fundamentally sound. In all, the Army built eight nuclear reactors. Fully half of them operated reliably for ten years or more. The explosion at SL-1 may have planted the seed of doubt in

U.S. ARMY NUCLEAR REACTORS			
REACTOR	LOCATION	CRITICAL	DECOMMISSIONED
SM-1	Fort Belvoir, Virginia	April 8, 1957	1973
SL-1	NRTS Idaho	August 11, 1958	1961
PM-2A	Camp Century, Greenland	October 3, 1960	1962
ML-1	NRTS Idaho	March 30, 1961	1966
PM-1	Sundance, Wyoming	February 25, 1962	1968
PM-3A	McMurdo Sound, Antarctica	March 3, 1962	1972
SM-1A	Fort Greely, Alaska	March 13, 1962	1972
MH-1A	The *Sturgis*	January 24, 1967	1977

some minds, however, inside the service and out, that perhaps the Army really shouldn't be in the expensive, dangerous business of building and operating nuclear power plants.

A year after SL-1 exploded, PM-2A at Camp Century came to the end of its life, and couldn't find a new home. It set the pattern for all the Army reactors, as they drifted toward the end, useful but unloved. The program dwindled and it budgets shrank. The Army's longest-lasting reactor, and the symbol of its program, was SM-1 in Fort Belvoir. It was decommissioned in May 1973, its ceremonial final shutdown performed by some of the original crewmen. In all, the reactor operated for sixteen years. The Army was always the most ecumenical of the military nuclear power programs, and the 813 alumni of SM-1 training course reflected that. The plant trained 474 men from the Army, 233 from the Navy, 101 from the Air Force, and five civilians.

By 1974, only the power barge *Sturgis* remained operational in the Army program. In the late 1950s, when the Army nuclear program was born, its generals imagined themselves manning a chain of Arctic radar stations, democracy's last line of defense against Soviet bombers. Twenty years later, its only reactor was reliably generating power in the heat of the Canal Zone of Panama. The Army was by then bogged down in the jungles of Vietnam, so devoting huge amounts of money to nuclear power stations designed for the North Pole seemed increasingly frivolous. The program ended with a whimper when the *Sturgis* left its longtime home in Panama in 1977. Among other reasons, the ship was seen as too attractive a target for Panamanian radicals. On its way back to Fort Belvoir, a severe storm battered the vessel. With some minor deliberation, the Army decided not to repair its only remaining nuclear reactor. The *Sturgis* was decommissioned and the Army nuclear power program became the Facilities Engineering Support Agency, an organization focused on providing generators of the conventional sort.

In 1975, Army Lieutenant General James Lampert, retired, wrote a polite letter declining to attend a ceremony in honor of the decommissioned SM-1. "I have very strong memories of those days," he wrote, "and would have enjoyed being with you." But he had been retired from the military for three years at that point, and was busy in his post as the chief fund-raiser at MIT. Lampert was undoubtedly proud of the Army nuclear power program and his contributions to it, but neither he nor the service he loved had been transformed by the experience. Lampert and the Army nuclear power program had succeeded quietly, while Keirn and the Air Force's nuclear airplane had failed loudly, but neither man ever came close to the accomplishments of Rickover. The Army and the Air Force had selected competent officers with the requisite credentials to head their programs. The Navy, in contrast, had quite unintentionally engaged the services of a revolutionary.

The nuclear airplane died with much more drama than the Army program, with Kennedy's sudden cancellation and the dire warnings that the United States would soon be outmatched by a fleet of Soviet nuclear jets. Even in its death, the Air Force program would continue to attract a more passionate defense than the more successful Army program ever had. People who devoted years to the program are understandably reluctant to say that it was doomed from the start, burdened with impossible technological challenges and a lack of visionary leaders. Bob Drexler, who worked as a mechanical engineer for General Electric on the nuclear airplane project from 1955 until the end, says he remains "totally convinced the airplane could have been made." When asked about the contrast between Rickover's fleet and the Air Force's inability to come up with a single flying prototype, Drexler echoes General Keirn's position from all those decades ago, that the ANP's project was "far more difficult."

Much like the defenders of the war in Vietnam, the running theme of the nuclear airplane enthusiasts was that the plane would have flown, had only the politicians stayed out of the way and not provided what Donald Keirn called "a lot of political by-play that would take a book to relate." Even Susan Stacy, in her otherwise clear-eyed history of the Idaho site, titles her chapter on the nuclear airplane "The Triumph of Political Gravity over Nuclear Flight." Stacy writes in closing: "The vacant TAN facilities went up for rent, a testimonial that the NRTS, no matter how brilliant its scientists and engineers, could not control its destiny when the political winds of Washington blew across the desert." The nuclear airplane had, in fact, enjoyed lavish political support, to the tune of a billion dollars over more than a decade during which the program created no visible signs of progress. Those who would continue the project had mastered the art of Cold War scare tactics, the well-timed article and unsubstantiated reports of Soviet planes with unlimited endurance. The Air Force and GE had only politics to thank for keeping the program alive for so long. Kennedy's cancellation of the program was at once an act of common sense and political courage.

• • •

Rickover continued to outlast his opponents, and most of those men he had put in command of his ships. He reached the Navy's mandatory retirement age in 1962, at the age of sixty-two, and once again showed that he could be flexible, at least when interpreting the Navy regulations as they applied to him. He received a special dispensation from Secretary of the Navy John B. Connally to stay on for another year. Rickover's carefully constructed network of friends in Congress, combined with the general impression, also carefully cultivated, that he was indispensable, made a string of future presidential administrations do the same. Johnson, Nixon, Ford, Carter: all would continue the pattern, even promoting Rickover to full admiral and four stars at the age of seventy-three. Scoop Jackson, one of Rickover's oldest and most reliable friends in Congress, wrote the resolution for that promotion. That same year, Rickover received from Congress what must have been an even more satisfying honor. Both houses recommended to the Naval Academy, Rickover's oldest nemesis, that their new engineering building be named Rickover Hall. It was emblematic of Rickover's career: the congressmen who worshipped Rickover ordering the reluctant Navy brass to honor him. A bust for the building's foyer was crafted using bronze from the *Nautilus.* The nose is now shiny from the passing hands of thousands of midshipman engineers looking to Rickover for luck. The bust appears to depict the admiral in civilian clothes.

Rickover had witnessed, and could take much credit for, the ascendance of nuclear power as a viable means of producing electricity for the public. While serving as an officer in the U.S. Navy he saw nuclear energy change from a fantasy of the true submarine into a nearly commonplace fact of life. Rickover always worried, however, that those plants not directly under his control might someday again reveal the dark side of unlimited power. It had happened at SL-1. On March 28, 1979, in the heart of Pennsylvania, it happened again.

Three Mile Island is a sliver of land in the Susquehanna River, so named because it is three miles downriver of Middletown, Pennsylvania. About half the length of the island was completely covered with two nuclear power plants: TMI-1, operational since 1974, and TMI-2, operational for almost exactly one year at the time of the accident. On that night, TMI-1 was shut down for refueling, but TMI-2 was operating at close to full power, generating 900 megawatts of electricity. Both plants were pressurized water reactors, larger-scale versions of the plants Rickover had pioneered. And that was just one imprint of the Navy visible at Three Mile Island. Nearly all the operators were alumni of Rickover's program, including shift supervisor William Zewe and shift foreman Fred Scheimann. Craig Faust and Edward Frederick, also nuclear Navy veterans, were standing watch in the control room when things began to unravel.

At 4:00 AM, a condensate pump shut down unexpectedly. The pumps fed water to the higher-powered feed pumps, which in turn forced water into the steam generators. When the condensate pump tripped off, the associated feed pump turned off automatically. With no water going into the steam generators, the reactor had lost its method of removing heat. The reactor's protection system functioned flawlessly, and automatically scrammed the reactor.

With all the rods at the bottom of the core, the chain reaction inside the reactor's uranium fuel stopped instantly. However, a number of nuclear reactions still took place in the core, a normal condition upon shutdown that generated "decay heat." The temperature of the primary system rose, and with it, so did the pressure.

At this point, another safety mechanism built into the plant functioned exactly as designed. When the pressure rose to a preset point, a relief valve lifted, discharging primary fluid in order to keep the plant from reaching an unsafely high pressure. From the time the condensate pump first tripped until the relief valve lifted, just three seconds passed.

When the pressure lowered to a safe point, after about seven seconds, the valve was supposed to close. It did not. The unexpected shutdown of the condensate pump at 4:00 AM was the minor problem, easily overcome, that initiated the sequence of events at Three Mile Island. The failure of the relief valve to close was the major malfunction that turned it into a disaster.

With the relief valve stuck open, primary fluid gushed out of the plant, steadily lowering the level of liquid in the core. What's worse, the operators, trying to assess hundreds of blinking alarms and screeching sirens in the control room, failed to interpret what was happening correctly. They feared that the plant was completely filling with water, becoming "solid," a dangerous state in which reactor pressure becomes almost impossible to control. In fact the exact opposite was happening—the plant was hemorrhaging water. After two minutes, yet another of the plant's automatic protective systems functioned appropriately. Correctly sensing the lowering water level in the core, an emergency cooling system turned on and shot a thousand gallons of water per minute into the core.

After two minutes the operators, fearful that it would fill the primary system completely, shut off the emergency cooling system. It was their most grievous error. With the emergency cooling system disabled, water from the reactor continued to exit the plant via the open relief valve until the fuel elements inside the reactor were uncovered. With nothing to remove their heat, temperatures soared, and the fuel melted.

Over the next several days, a nervous nation grappled with the disaster in central Pennsylvania. Catholic priests in the area granted general absolution to tearful congregations, and a voluntary evacuation of pregnant women and young children was announced. Tensions reached a pinnacle when a hydrogen bubble was discovered inside the containment building. Some theorized that the hydrogen would explode, breach the containment, and make central Pennsylvania uninhabitable for centuries.

President Jimmy Carter, accompanied by his wife Rosalynn, took a tour of Three Mile Island on April 1, 1979. His goal was to show personal concern for the citizens of the area, and at the same time mitigate the panic by demonstrating that he was willing to personally visit the site. President and Mrs. Carter toured the control room and the surrounding area wearing yellow booties over their shoes to prevent the spread of contamination. With that, Jimmy Carter became one of very few people on earth to have toured the sites of two nuclear accidents, going back to his experience at Chalk River in Canada in 1952.

Carter's old taskmaster from his days in the Navy was naturally asked for his opinion about the accident. Before Congress, Rickover observed that the men at Three Mile Island were largely Navy-trained. And yet, "the thing wrong at Three Mile Island was not the design of the plant. It was the lack of supervision and carelessness in operation." Rickover explained why Shippingport, "his" civilian plant, had run twenty-five years without an incident in that same state:

> *Because I have my representative sitting in the control room every minute that reactor is operating. He sees to it that the people from Duquesne Light Company who operate it do their jobs properly. He watches them. They are not allowed to talk with each other except on official business. If he sees one of the operators talking to another and it's not business, he tells them to stop. If they don't stop, he shuts down the plant, and we have shut it down twice, because I maintain control of my reactors.*

In short, the problem with Three Mile Island was that Rickover was not in charge of it.

Three Mile Island became synonymous with nuclear disaster, and even the unaffected plant, TMI-1, couldn't be restarted because of public pressure on regulators to keep it closed. After four years, not knowing where else to turn, the utility that ran Three Mile Island made a $380,000 donation to Rickover's charitable foundation in return for an evaluation by the admiral of the plant's overall fitness. Rickover made his report in 1983, stating tersely that management had the "competence and integrity" to operate the plant safely. The plant was restarted on October 2, 1985, more than six years after the accident. It is still running today, and has become a regular winner of industry awards for safety and reliability.

At least one person close to him bitterly disputed the company's use of Admiral Rickover, eighty-three years old at the time he was commissioned to investigate Three Mile Island. The day after the admiral's death, Robert Rickover, his son, told the *New York Times* that his father had been "ruthlessly exploited at a time when he was not mentally competent."

The head of Rickover's foundation, Joann DiGennaro, disputed that assertion, and knowingly hinted that the son's protest might be more a statement on the relationship between father and son. "I don't know how to respond to a son who says that," DiGennaro said, "except to say that I wish he had seen his father more often."

When TMI-2 could finally be accessed, its fuel was removed. The NRC Fact Sheet on the accident says that fuel and debris were "shipped off-site to a Department of Energy Facility." Even though the distance was great, there was really only one place in the nation that could house the twisted radioactive remnants of a nuclear accident, the same place that had received the remnants of SL-1 in 1961: Test Area North, the hot shop that would have been home to the atomic airplane. The radioactive debris from Pennsylvania was shipped by rail, forty-nine casks in twenty-two shipments, two thousand miles, to Idaho.

• • •

Rickover was planning to live forever, it seemed, and to hold on to his job just as long. He spent most of his final years of active duty in an absolute war with General Dynamics, which had become parent company of Electric Boat, pointing out to all who would listen that the contractor was deliberately underbidding Navy contracts only to bilk the taxpayers later with massive cost overruns. Rickover declared that this had become business as usual not just at General Dynamics, but in the whole world of defense contracting. It was embarrassing to both the corporate executives and the politicians who sponsored them, but no politician had the will to take on Rickover. After all, Rickover's results spoke for themselves. While the Air Force and Army programs were little-known historical footnotes, nuclear ships had become keystones of both naval tactics and the strategic doctrine of the United States. At the end of his career, Rickover could boast of 121 nuclear submarines, three carriers, and nine other nuclear-powered surface warships, with a score more authorized or under construction. And Rickover's perfectionism had resulted in an extraordinary safety record: zero nuclear accidents in the three decades since the launch of the *Nautilus.* By Rickover's own estimate before Congress, that added up to 2,300 years of reactor operation without an incident.

But even Rickover couldn't rule his kingdom forever. In 1981, the newly elected Ronald Reagan, eleven years younger than Rickover, finally determined that the admiral's sixty-three years in (and out of) uniform were sufficient. The president detailed his secretary of defense, Caspar Weinberger, to tell him the news, and finally, at the age of eighty-two, Rickover was forced out of the military. While Reagan declined to attend, a retirement party for the reluctant retiree included three ex-presidents as guests: Carter, Ford, and Nixon. (It was at this event that Bob Dole issued his famous description of the three men as "See no evil, hear no evil, and evil.") When the hosts requested a military band, the Pentagon flatly turned them down. Rickover's friends, his saviors on many occasions, had always been elected officials. But he had spent a lifetime antagonizing the brass and government contractors, and they were only just beginning to exact some measure of revenge.

Although unable to save his job, Rickover's friends in Congress called him to testify one last time. Rickover spent his last day in uniform, January 31, 1982, not in the conning tower of a submarine or in the control room of a reactor, but in a hearing room on Capitol Hill, where his true genius had always been most apparent.

The hearing began with the predictable plaudits. Senator William Proxmire of Wisconsin called Rickover "a National Treasure." Rickover's dependable ally, Henry "Scoop" Jackson, called him a "breath of fresh air" and called his forced retirement "the nation's loss." Congressman Henry Reuss of Wisconsin summed it all up with "We love you, Admiral," while the admiral pretended to be embarrassed by the praise.

Rickover then began his lengthy statement, a statement that was in many ways typical of the many he had made since his first appearance before Congress more than three decades earlier. There were obscure classical references: "The Moor has done his duty, and the Moor may go." Ancient legal references: "Ever since the famous Santa Clara County v. Southern Pacific Railroad case of 1886 . . ." He took a jab at those who would have him retire because he was too old: "The Navy medical staff has certified that I am fit in all respects for continued active duty." And, of course, he took a swipe at the Naval Academy with a familiar complaint about its overemphasis on athletics: "The only point for wrestling that I know is maybe they know how to wrestle with girls. I see no other purpose." His most voluminous criticism was aimed at those corporations who profited from his program.

A preoccupation with the so called bottom line of profit and loss statements, coupled with a lust for expansion, is creating

> *an environment in which fewer businessmen honor traditional values; where responsibility is increasingly disassociated from the exercise of power; where skill in financial manipulation is valued more than actual knowledge and experience in the business; where attention and effort is directed mostly to short-term considerations, regardless of longer-range consequences.*

Rickover singled out one contractor, along with those politicians who seemed reluctant to look too closely at their practice of submitting claims to the Navy after a contract had been signed and work was under way:

> *After investigating General Dynamics, our biggest defense contractor, for four years, the Department of Justice recently announced they could find no evidence of criminal intent, although the claims were almost five times what the Navy actually owed.*

Rickover's testimony took a curious turn when Senator Proxmire asked him about the long-term prospects for civilian nuclear energy. Rickover responded by explaining how nuclear power created radiation, something that had had to be reduced millions of years ago on Earth to allow life even to exist. To create radiation, Rickover concluded, was in some ways to go against nature. While Rickover had long been recognized as the nation's nuclear patriarch, those closest to him often discerned this ambivalence about nuclear energy. He had always seen nuclear power as something worthwhile only if the survival of the nation depended upon it, and, even under those circumstances, something that required diligence of religious intensity. These feelings came out in his final congressional testimony, and it startled many in the room, even those like Senator Proxmire who were accustomed to Rickover's trademark churlishness.

> Admiral Rickover: *I do not believe that nuclear power is worth it, if it creates radiation. Then you might ask me, why do I have nuclear powered ships? That's a necessary evil. I would sink them all. Have I given you an answer to your question?*
>
> Senator Proxmire: *You've certainly given me a surprising answer. I didn't expect it and it's very logical.*
>
> Admiral Rickover: *Why wouldn't you expect it?*
>
> Senator Proxmire: *Well, I hadn't felt that somebody who's been as close to nuclear power as you have and who's been so expert in it and advanced it so greatly would point out that, as you say, it destroys life.*
>
> Admiral Rickover: *I'm not proud . . .*
>
> Senator Proxmire: *Without eliminating it or reducing it many, many years ago, we couldn't have had life on earth. It's fascinating.*
>
> Admiral Rickover: *I'm not proud of the part I've played in it. I did it because it was necessary for the safety of this country. That's why I'm such a great exponent of stopping this whole nonsense of war.*

With Rickover a civilian, his many enemies finally had a chance to act. In 1984, a disgraced Electric Boat executive, P. Takis Veliotis, actually fled the United States to avoid prosecution for the millions of tax dollars his company had bilked from the government, corruption Rickover had relentlessly pursued until his last day on active duty. From Greece, Veliotis sent a *Washington Post* reporter a carefully tabulated file of gifts Rickover had received over the years from the contractor, mostly trinkets of the tie-clip and ship model variety. Rickover had accepted many car rides from Electric Boat over the years; Electric Boat had documented them all and called them gifts. There were larger gifts, too, ones that were harder to explain, gifts Rickover probably should have known better than to accept: diamond earrings and a jade pendant, for example, valued at $1,125. The jewelry was symptomatic of one of the more troubling aspects of Rickover's personality—his view that he was above the rules, whether for retirement age, wearing a uniform, or accepting expensive gifts from contractors. It was a glaring contradiction in a man who demanded unwavering procedural compliance aboard his ships.

Still, Rickover was no more likely to favor a contractor for such a thing than he was to wear a jade pendant. Electric Boat, in particular, could never have plausibly argued that Rickover somehow took it easy on them because of the gifts. Nonetheless, Rickover had violated Navy regulations by accepting anything and creating the appearance of impropriety. The total value of the gifts, released gleefully by the Navy brass in a thirty-two-page report, was $68,703.

Rickover never claimed that he hadn't taken the gifts. Instead, he challenged his accusers to find one example of when the gifts had ever influenced him to treat Electric Boat or its parent company, General Dynamics, with anything other than his normal vigor. By all appearances, General Dynamics released the details *because* Rickover was so diligent in exposing their cost overruns. It didn't matter. Cementing the decision to kick Rickover out the door, Secretary of the Navy John Lehman put a letter of reprimand into Rickover's thick personnel file.

Rickover would grant one more interview before his death, to *60 Minutes* in 1984. In the interview, he again railed against General Dynamics and defended himself against the charges that he had somehow behaved unethically. Also in that interview, he made one of his few public statements ever about anti-Semitism during his years at the academy. Diane Sawyer asked him why he perhaps had a harder time than others during his years in Annapolis. He answered, "Because I was Jewish. They didn't have any—very rare for a Jew to go to the Naval Academy."

During that interview Rickover also directed part of his grouchiness toward the interviewer herself, and one senses that she would almost have been disappointed if it hadn't happened. He told Sawyer, "No, I never have thought I was smart. I thought the people I dealt with were as dumb, were dumb, including you."

The well-prepared Sawyer responded that she was in good company, because Rickover had said almost the same words in an interview decades before with Edward R. Murrow.

The charges of misconduct bothered Rickover until the very end. As he told his old friend Ted Rockwell on his deathbed, the letter of reprimand was "the last word in my personnel jacket, the first piece of paper you see when you open it up."

Rockwell responded, "Admiral, your record will speak for itself."

Rickover would have to let his record speak for itself, because he wrote no memoir. He told AEC historian Francis Duncan that most autobiographies were "self serving and of limited value." He also energetically refused to cooperate with the one serious biography of him written during his lifetime: Norman Polmar and Thomas Allen's *Rickover: Controversy and Genius*. Rickover wrote to the chairman of Simon & Schuster, the book's publisher, warning, "should there be any inaccuracies in the book that affect me personally, your company could be liable to a suit." Rickover became even more enraged when he got word that Peter Douglas, son of Kirk Douglas, was interested in making a film version of the biography. In Douglas's words, Rickover "made it clear that he will do everything legally possible to obstruct the project." Rickover succeeded in killing the movie project, but not the book, which was published in 1982 almost concurrently with his forced retirement.

While he may have found memoirs of limited value, Rickover did write two book-length works of history that were informative in their own way about their author. First came *Eminent Americans: Namesakes of the Polaris Submarine Fleet*, published in 1972. The book contained elegant biographies of each man who had a Polaris submarine named in his honor, forty-one in all. It was an eclectic group, one that included men who actually took up arms against the United States government (Tecumseh, Stonewall Jackson, and Robert E. Lee), a king who never set foot on U.S. soil (Kamehameha), and a humorist (Will Rogers). Of all the essays Rickover wrote, Thomas Edison's was perhaps the most revealing:

> *A technical man himself, he understood that inventiveness—whether individual or collective—flourishes best in an atmosphere of freedom, where the productive men are protected against interference by nontechnical "administrators." After Edison, invention became a virtual monopoly of huge private and public bureaucracies managed by nontechnical "organization men," overfond of administrative charts and regulations.*

In 1976, Rickover published *How the Battleship Maine Was Destroyed*, a brief but thorough investigation of the 1898 naval disaster that started the Spanish-American War and set the United States on the course of becoming a global power. Rickover disputed the conventional explanation that a Spanish mine blew up the ship, but instead found an accidental explosion far more likely. There are several telling passages in the book, such as when he finds it suspect that the *Maine*'s commanding officer, Captain Charles D. Sigsbee, had previously been in command of ships that inspectors twice found "dirty." Rickover also wrote this damning passage about Sigsbee's testimony before the Court of Inquiry, in which he took the man to task for a variety of inadequacies. It is, in many ways, a one-paragraph dissertation on what Rickover believed a naval officer should not be, a catalog of all the sins he had spent a lifetime trying to eradicate.

> *From his testimony emerges the portrait of an individual who was unfamiliar with his ship. He might have been a good seaman and a brave man, but perhaps also the victim of the new technology which was transforming the Navy. He might not have understood the complexities of the ship he commanded. He might have suffered from the division in the Navy that separated line officers from the shipboard engineers. Many line officers looked down upon engineering. The vagueness and uncertainty in his testimony might stem from a belief that giving an order was tantamount to its execution. Whatever the reasons, he appears to have been isolated from the day-to-day routine.*

Rickover would live to see one more nuclear disaster, this one far worse than either SL-1 or Three Mile Island. On April 26, 1986, the Soviet power plant at Chernobyl exploded, instantly killing around fifty people. The total number of fatalities caused by the explosion and its massive plume of radiation will be debated forever, but it is certainly in the thousands. The accident at Chernobyl, while exponentially more severe than SL-1, bore many similarities. The explosion occurred during maintenance on a shutdown reactor, performed by an unsupervised, skeleton crew, which mistakenly withdrew control rods, at 1:23 in the morning. If control rod maintenance after 9:00 PM had been somehow prohibited by international law, the accidents at SL-1 and Chernobyl might both have been averted.

While Rickover was still alive at the time, he was by then too unhealthy to venture forth a comment or a pointed opinion about why another of the world's reactors that didn't belong to him had gone haywire. On July 4, 1985, Rickover suffered a stroke. That prevented him from attending the ceremony as the *Nautilus,* her reactor having been removed at Mare Island Naval Shipyard in California, was towed into Groton for the last time. She would become a museum after serving the nation for twenty-six years.

Hyman G. Rickover died on July 8, 1986.

EPILOGUE

The Donald C. Cook Nuclear Power Plant rises like a Norman castle along the pristine eastern shore of Lake Michigan, bounded by water on one side and dense forest on the other, a beige band of sand dunes separating the blue from the green. Forty years of controlled access and tight security around the plant's 650 acres have made the area something unexpected: a place of pristine natural beauty.

Driving up the hill to the visitors center, the plant comes in view: two huge concrete domes, barricades, high fences, and monumental transformers and cables that lead Cook's electricity to the North American power grid. Cook lacks what is to many a fundamental symbol of nuclear power: a set of hyperbolic cooling towers. Bill Schalk, communications manager, displays a nimble sense of pop culture when he theorizes to visitors about why the towers have become so emblematic of his industry: "Three Mile Island and *The Simpsons*." Cooling towers are just structures that remove heat from water, and are not necessarily a characteristic of nuclear power. (Cook, like a nuclear submarine, instead uses the nearby large natural body of water as its ultimate heat sink.) In contrast, another power plant near Cook, in Michigan City, Indiana, is a coal-burning facility, but is often mistaken for a nuclear plant because of the huge hyperbolic cooling tower that looms over it, steam billowing from it impressively around the clock. In photographs of nuclear power plants with towers, they are inevitably the most intriguing and sinister feature in the frame, but in fact there is nothing the least bit radioactive inside those futuristic curves. The reactor core typically resides inside a smaller, much less dramatic building next door.

Next to the power plant, inside a lodgelike visitors center, animated displays illustrate the cleanliness, security, and remarkable safety record of the nuclear power industry. It is more dangerous, visitors learn, to be a real estate agent than it is to work in a nuclear power plant. The 1979 accident at Three Mile Island is dealt with forthrightly, as it is in most pronuclear literature. The incident is the industry's answer to the job interview question, "describe for me a challenge that you've overcome," a story of lessons learned, obstacles conquered, and the stronger, safer industry that resulted. And after all, it is often pointed out, no one died at Three Mile Island. Bill Schalk accepts the role of industry spokesman with enthusiasm, and is nearly giddy when he describes the prospects for nuclear power in the United States. The nation, he believes, is finally beginning to recognize what they have known at Cook for nearly forty years: atomic energy is good for America.

Between the visitors center at Cook and Lake Michigan is a spacious patio, well populated with pergolas and picnic tables. Prior to September 11, 2001, the Cook Plant was a focus of community life in rural southwestern Michigan. The plaza regularly hosted thousands for weekend craft shows and the like, before tighter security made it impossible to allow so many unscreened people within a few dozen feet of a nuclear power plant. Scheduled tours are still allowed, but the loss of this openness is something Cook executives clearly regret—they pride themselves on the solid community relations they have built during the plant's three decades of operations. In the lobby of the visitors center, and displayed more prominently than any of their numerous industry awards, hangs a certificate from the local *Herald-*

Palladium giving the Cook Plant a 2003 Reader's Choice Award: "Best Place to Take the Kids."

While spokesmen like Bill Schalk studiously avoid the word, the American nuclear power industry is on the verge of a boom. The expensive and time-consuming licensing procedure for new plants has been vastly simplified. When the commercial industry was born, utilities had to get their plans approved prior to construction, and then apply for a second license before beginning operations. This cumbersome process was replaced by the one-step "Combined Operating License" in 1989. Several companies that make nuclear reactors, including General Electric and Westinghouse, have gone to the trouble and expense of having their newest plant designs "preapproved" by the Nuclear Regulatory Commission in a way that could subtract years from the building and permitting process. As one further measure of financial security to utility companies, the federal government in 2006 authorized a kind of risk insurance for the next six new plants in the United States. Enterprising utility companies with faith in nuclear power will be compensated for regulatory delays not of their own making, up to $500 million apiece.

The NRC in its bulletin "Expected New Nuclear Power Plant Applications," updated on October 11, 2007, lists an expected total of five new applications in 2007, fourteen in 2008, and two in 2009, totaling twenty-one applications for thirty-two reactors. NRG Energy became the first utility company to apply for a license since TMI in September 2007, when it applied to the NRC for permission to begin building two new plants in Matagorda County, Texas, along the Gulf Coast. The location is already home to two nuclear reactors, and while the application is generating some protest, local support is strong. Nuclear power has been safely providing electricity and good jobs in the rural area for nearly two decades.

Political conservatives have always been nuclear power's natural allies, fans of both energy independence and the huge corporations who would benefit from any new investment into the industry. They have some unexpected new allies on the left. Several name-brand environmentalists have broken ranks and taken up the cause of nuclear power, including Gaia theorist James Lovelock, *Whole Earth Catalog* editor Stewart Brand, and Greenpeace cofounder Patrick Moore. They argue that a dogmatic fear of nuclear power, and technology in general, has made some of their fellow environmentalists irrational on the subject. Stewart Brand in 2005 called the environmentalist aversion to nuclear energy "quasi-religious." Furthermore, the dangers of global warming have become so great that nuclear power, an emission-free alternative, deserves new, serious consideration. James Lovelock pioneered "Gaia Theory," a school of thought that holds that the planet Earth acts as a single, living organism. He shocked many of his colleagues with an editorial in Britain's *Independent* in 2004, summing up his pronuclear argument in dramatic fashion: "We have no time to experiment with visionary energy sources; civilisation is in imminent danger and has to use nuclear—the one safe, available, energy source—now or suffer the pain soon to be inflicted by our outraged planet."

The hopes of the new pronuclear coalition were symbolized by a visit President George W. Bush made to Calvert Cliffs Nuclear Power Plant in Maryland, on June 22, 2005. It was the first visit by a sitting president to a nuclear plant since President Carter's dramatic walk in the yellow booties at Three Mile Island in 1979, and the contrast between the two visits marked how far the industry in the United States had traveled since that low point. In his speech, Bush was unequivocally supportive of nuclear power, "the one energy source that is completely domestic, plentiful in quantity, environmentally friendly, and able to generate massive amounts of electricity." Two days later, the U.S. president with the most personal expertise in nuclear energy visited another nuclear plant. Jimmy Carter, in the area for a Habitat for Humanity project in Benton Harbor, Michigan, toured Donald C. Cook Nuclear Power Plant.

Cook is a typical American nuclear power plant in almost every way. The capacity of its two units are 1,016 and 1,077 megawatts, both close to the median for all 104 licensed U.S. plants: 993 megawatts. The exact conversion factors are debatable, but it is often stated that 1 megawatt is enough electricity to supply a thousand U.S. homes. The two Cook plants combined, 2,093 megawatts, supply enough electricity for approximately 2.1 million homes—enough to supply all the homes in Chicago and Houston combined. It's a staggering amount of power concentrated in one location, so much so that after an extended shutdown of both Cook plants ended in 2000, the trade magazine *Coal Week* published an article speculating that the sudden reduction in demand might negatively affect the price of coal worldwide.

It isn't just capacity that makes Cook typical. Cook 1 began commercial operation on August 27, 1975; Cook 2, July 1, 1978. The median start date for all U.S. plants is February 16, 1979, during what could be called the heyday of the nuclear power industry. Both Cook reactors were built by Westinghouse, the largest supplier in the U.S. industry, manufacturer of forty-eight of the nation's reactors. General Electric is in second place, with thirty-five. One of the reasons both GE and Westinghouse willingly submitted to Rickover's tyranny was that each hoped it would give them a lasting head start in the emerging civilian nuclear power business. It appears that this calculation was correct. Like the vast majority of the nation's reactors, both Cook reactors fall into the category of pressurized water reactors, or PWRs, the model of reactor pioneered and advocated by Rickover.

This is part of the legacy of SL-1, a boiling water plant. But perhaps the most lasting legacy of SL-1 is more subtle. The demise of the Army program (along with the Air Force's) allowed the Navy complete hegemony in the nation's nuclear power plants, ensuring that Navy standards would become industry standards. Those standards, at their philosophical core, remain the standards of Admiral Hyman Rickover. And while the Cook plant is a robustly civilian, profit-making enterprise, wholly owned by the publicly traded corporation American Electric Power, the stamp of the U.S. Navy at Cook is everywhere. And it is not just because the plant is poised on the edge of Lake Michigan like a ship of war, although that does contribute to the overall nautical feeling, as gulls caw in the background and buoys ring their lonely bells. Ex-Navy nukes populate every department at Cook, as they do at every American nuclear plant, as they do in all the regulatory agencies that monitor them. The Navy every year unleashes hundreds of its highly trained, experienced reactor operators into the civilian world, where they are welcomed by the civilian industry. The Nuclear Energy Institute, the industry's primary trade group, is headed by Skip Bowman—he was on the podium with President Bush at Calvert Cliffs. Bowman is a former admiral and top Navy nuke, one of the heirs to Rickover's throne.

It is a well-known attribute of the civilian nuclear industry, this cultural link to the Navy, and it is often commented upon in writing about the business. The industry doesn't mind the connection, as it links their performance and procedures to the Navy's perfect safety record over fifty years and hundreds of reactors. Even the civilian hierarchy retains some resemblance to its military counterpart. Utilities with multiple nuclear plants refer to them as "fleets," and many top executives in the industry have a ship's plaque and officer's ribbons displayed somewhere in their offices. Even twenty-five years after his retirement, many of these executives have stories about their interviews with Rickover, the day he personally deemed them worthy of tending a nuclear reactor.

Out in the engine rooms, where the actual sweaty work of power generation is accomplished, the men are more likely to have tattoos and enlisted men's service records. While the movie *The China Syndrome* is famously lax in its scientific depiction of nuclear power (Jack Lemmon's character, Jack Godell, threatens late in the movie to "flood the containment with radiation!"), it does get this cultural detail correct. After the initial incident, Ted Spindler, played by Wilford Brimley, laments his chances in the board of inquiry, certain that he will be singled out because he is not a member of the fraternity: "You and the rest of the hot-shot Navy boys have credentials." *The China Syndrome* cemented its place in pop culture by coming out twelve days before the accident at Three Mile Island. Since then, it is often pointed out, not a single new nuclear plant has been ordered in the United States.

What is less often pointed out is that a great many plants have opened for business in this period. It takes a long time to build a nuclear power plant, and many plants ordered before TMI came on line not all that long ago. Of the nation's 104 licensed plants, exactly half began commercial operations *after* the accident at Three Mile Island. The most recent is Watts Bar 1 in Tennessee, which began commercial operation on May 5, 1996. (The oldest is Oyster Creek, in Ocean County, New Jersey, operational since 1969.) At the time of Three Mile Island, nuclear power generated 12.5 percent of the nation's electricity. Ever since then, in the wake of what was supposedly an industry-killing public relations disaster, the percentage has actually gone up. Today, nuclear power supplies 19.4 percent of the nation's electricity.

The growth of nuclear power has been even more pronounced in other countries. According to the International Atomic Energy Agency, the group founded in the wake of Eisenhower's "Atoms for Peace" speech, eighteen countries produce a higher percentage of their electricity from nuclear power than the United States. Four countries get more than half their electricity from nuclear power: Belgium, Slovakia, Lithuania, and the world's nuclear leader, France. France has fifty-nine nuclear plants, including the four largest in the world, which generate a total of 78 percent of France's electrical power. This massive commitment to nuclear power was a deliberate strategy undertaken by France in the 1970s to ensure its energy independence, noting they had a wealth of engineering talent but a paucity of indigenous fossil fuels. Today France is the largest electricity exporter in the world, and has the lowest cost of electricity in Europe.

Even with the surprising success of nuclear power in the United States, coal still generates the lion's share of the nation's electricity at about 49.9 percent. And while the worst-case scenarios attached to nuclear power can provide terrifying hypothetical fatalities, the hunger for coal kills real people, dependably, every year. Since Three Mile Island, according to the National Institute for Occupational Safety and Health, 130 coal miners have died in "disasters," those accidents that killed five or more people at a time. The list of disasters since 1839 goes on for fourteen pages, and tersely describes how 13,805 American coal miners died: explosion, fire, haulage, and "inundation" have all claimed their share of victims. The worst disaster was in 1907, in Monongah, West Virginia, when 362 miners died in a series of explosions. As recently as January 2, 2006, twelve men died at the Sago Mine, again in West Virginia, again in an explosion. After ninety-nine years, the hazards and locations of coal mines have changed little.

And these deaths obviously do not include black lung, the impact of global warming, acid rain, smog, or any of the other side effects that result from this country burning a billion or so tons of coal every year. While the benefits of nuclear power are often repeated by industry spokespeople, they rarely articulate the hazards of coal dependence. The companies that own nuclear power plants are by and large utilities that also own profitable coal-burning plants. They are unlikely to start denigrating their biggest fuel source, no matter how enthusiastically they advocate nuclear power.

Another reason the nuclear industry is reluctant to tell its own story might be yet another vestigial remnant of its naval heritage. While there are nuclear-powered surface ships, most notably the aircraft carriers, the vast majority of naval reactors, and Navy nuclear-trained people, come from submarines. This is a group that keeps its accomplishments secret. Secrecy is, in fact, a kind of cult within the "Silent Service," a well-established source of institutional pride. Within other specialties of the Navy, this attitude is often taken for arrogance—which is undoubtedly a component. It also originated with real security requirements, not surprisingly in an organization that grew up in the Cold War, and in a tactical environment where stealth was often a ship's only viable defense. The secrecy is also due to the submarine force being a community of engineers, not a community of PR professionals. The facts, and their accomplishments, these engineers believe, should speak for themselves.

Rickover is buried at Arlington National Cemetery. His gravestone contains the four stars of his rank, the dates of his birth and death, and the inscription "Father of the Nuclear Navy: 63 Years Active Duty." The grave is in one of the more exclusive sections of Arlington, within sight of Kennedy's eternal flame, a section filled with large personalized monuments, not the uniform rows of white marble markers that characterize the National Cemetery. Some of Rickover's immediate neighbors are Supreme Court justices: Thurgood Marshall, Harry Blackmun, and Oliver Wendell Holmes. While Rickover's section is filled with historic notables, however, it is also immediately adjacent to Section 31, and within sight of plot 472, the final resting place of Richard McKinley.

Fewer people visit McKinley's quiet grave than Rickover's, and most of those that come near it are probably there to see the adjacent gravestone of Dwight H. Johnson, who won the Medal of Honor in Vietnam only to be killed while robbing a liquor store in Detroit in 1971. Since McKinley was buried on federal property, it is not surprising that his interment generated a bit of government paperwork, a more robust version of the letter written by the Kingston, Michigan, cemetery's secretary to the AEC in 1961. It was one of thousands of government documents the SL-1 accident continues to emit, like radiation, an ever-decreasing but eternal stream:

> January 31, 1961
>
> Subject: Internment of Radioactive Remains
>
> To: Superintendent, Arlington National Cemetery
>
> 1. Radioactive Remains of SP4 Richard McKinley were interred at Arlington National Cemetery on 25 January 1961.
> 2. It is desired that the following remark be placed on the permanent record, DA Form 2122, Record of Internment: "Victim of nuclear accident. Body is contaminated with long-life radioactive isotopes. Under no circumstances will the body be moved from this location without prior approval of the Atomic Energy Commission in consultation with this headquarters."
>
> For the Commander:
>
> Leon S. Monroe, II
>
> 2nd Lt. AGC
>
> Assistant Adjutant General

While the warnings about McKinley's radioactivity are on file, no sign of his unusual death are on his marker. The irreducible laws of radiation and shielding, however, make it necessarily true that some infinitesimal amount of radiation still streams from McKinley's shattered body, crosses Sheridan Drive, and illuminates ever so slightly the grave of Admiral Rickover.

NOTES

12006 Wehman, George. *Preliminary Report of Fission Products Field Release Test—1*. U.S. Atomic Energy Commission, February 1959. (IDO-12006)

19017 *ABWR Quarterly Progress Report: SL-1 Operations and Evaluation*. Combustion Engineering, under contract for the U.S. Atomic Energy Commission, July 15, 1960. (IDO-19017)

19300 *SL-1 Reactor Accident on January 3, 1961: Interim Report*. Idaho Falls, Idaho: Combustion Engineering, under contract for the U.S. Atomic Energy Commission, May 15, 1961. (IDO-19300)

19301 *SL-1 Recovery Operations: January 3 Thru May 20, 1961*. Idaho Falls, Idaho: Combustion Engineering, under contract for the U.S. Atomic Energy Commission, June 30, 1961. (IDO-19301)

19302 *IDO Report on the Nuclear Incident at the SL-1 Reactor on January 3, 1961 at the National Reactor Testing Station*. Idaho Falls, Idaho: U.S. Atomic Energy Commission, January 1962. (IDO-19302)

19310 Islitzer, Norman. *The Role of Meteorology Following the Nuclear Accident in Southeast Idaho*. Idaho Falls, Idaho: U.S. Weather Bureau, May 1962. (IDO-19310)

19311 *Final Report of the SL-1 Recovery Operation*. Idaho Falls, Idaho: General Electric Co., June 27, 1962. (IDO-19311)

19313 *Additional Analysis of the SL-1 Excursion: Final Report of Progress July through October 1962*. Idaho Falls, Idaho: Flight Propulsion

Laboratory Department, General Electric Co., Nov. 21, 1962. (IDO-19313)

IFPR *Idaho Falls Post Register*

PROLOGUE: JANUARY 3, 1961

2 *six degrees below zero:* Actual temperature taken from the January 3, 1961, *Idaho Falls Post Register*. Many accounts of the night, such as that of Dr. George Voelz, would recall the night as being twenty degrees below zero or colder.

2 *SL-1 was the smallest: Thumbnail Sketch,* June 15, 1961, 1.

2 *200 kilowatts:* 19300, 1. There are some conflicting accounts of this, sometimes even in the same document. It appears SL-1's generator was rated at 300 kilowatts, but operated at 200 kilowatts. See 19300a, 1, 7.

4 *"The #2 crew member was struck":* 19300, 102.

CHAPTER ONE: THE USS *UNITED STATES*

11 *The* United States *by herself would cost $189 million:* contemporary estimate taken from the *Times-Herald* of Newport News, Va., April 18, 1949, 1, 12.

12 *"Controversy Still Rages on Capitol Hill Over Craft":* ibid., 1.

12 *"the facts in the matter are that the keel has been laid":* ibid., 12.

13 *"For your information the Marine Corps is the Navy's police force":* letter of President Harry S. Truman to Gordon L. McDonough, Aug. 29, 1950.

14 *"Truman's son-of-a-bitch":* Manchester, *American Caesar,* 532.

14 *"mentally ill":* Acheson, *Present at the Creation,* 374.

14 *"the only bull I know":* Barlow, *Revolt of the Admirals,* 174.

14 *"pathological condition":* Ferrell, ed., *Off the Record,* 192.

15 *"Admiral, the Navy is on its way out":* Heinl, *Victory at High Tide,* 6–7. This widely quoted passage originated here. Heinl states enigmatically, "The Johnson-Connolly quotation comes to me from a primary source who cannot for the present be identified" (269, n. 5).

16 *"The CVA-58 will probably carry": Time,* Oct. 11, 1948.

17 *"slow, expensive, very vulnerable":* Barlow, *Revolt of the Admirals,* 249.

18 *"The B-36 cannot hit precision targets":* ibid.

18 *"Why do we need a strong Navy":* ibid., 253.

18 *"I was not associated with Admiral Denfeld during the war":* ibid., 262.

19 *an incredible 280,000 shaft horsepower:* All design specs for the USS *United States* are taken from Friedman, *U.S. Aircraft Carriers.*

THE CADRE

21–27 Details are from the Miazga memos of Jan. 30, 1961, and July 25, 1962.

CHAPTER TWO: RICKOVER

29 *Hyman G. Rickover was born on January 27, 1900:* Polmar and Allen, *Rickover,* go into some detail to demonstrate that Rickover may have been born in 1898, and altered his birth date to increase his chances for admission to the Naval Academy. The 1900 birth date is what's normally given—either date makes Rickover older than the U.S. submarine force.

29 *Jewish schools:* Rockwell, *The Rickover Effect,* 20.

30 *"higher priority reasons to hate me":* Theodore Rockwell, interview with author.

30 *a perfect grade of 100:* Polmar and Allen, *Rickover,* 35.

31 *ranking 107th out of 540 midshipmen:* Rockwell, *The Rickover Effect,* 22.

32 *The top speed for the* S-48 *on the surface was 14.2 knots:* Tech specs for the *S-48* are taken from Friedman, *U.S. Submarines Through 1945.*

33 *decommissioned three separate times:* July 7, 1925, upon sinking during that New England storm; Sept. 16, 1935; and Aug. 29, 1945.

34 *would describe him as "gnomelike":* Zumwalt, *On Watch,* 88.

35 *"If you are an Admiral you should look like one"*: William McNally, "Admiral Rickover," http://www.nautilus571.com/rickover.htm.

36 *"painted over banana peels"*: Rockwell, *The Rickover Effect*, 32.

41 *"in view of the unusual and unpredictable hazards"*: W-31-109 Eng-52, contract between General Electric and the United States of America. Contract obtained from U.S. Department of Energy, Richland Operations Office, Richland, Wash. I have seen this contract cited as "W-31-109 Eng-32" but I now have a copy of the original, and it's "52."

41 *"We spent most time on a contract with General Electric"*: Lilienthal, *The Journals of David E. Lilienthal*, diary entry for Nov. 26, 1946.

41 *When it was required, he would write himself letters:* "The Man in Tempo 3," *Time*, Jan. 11, 1954.

42 *Rickover confidently declared:* Polmar and Allen, *Rickover*, 149.

42 *Rickover estimated his submarine:* ibid., 148.

43 *He spent his last moments as secretary of defense:* Truman's diary entry for Sept. 14, 1950, in Ferrall, ed., *Off the Record*, 193.

44 *MacArthur developed a detailed target list:* The whole incredible account of Operation Hudson Harbor is from Cumings, "Spring Thaw for Korea's Cold War?"

45 *"you can gauge our progress"*: "The Fastest Submarine," *Time*, Sept. 3, 1951.

45 *"elbowed their way into the family"*: Ruth Masters Rickover, *Pepper, Rice, and Elephants*, ix.

45 *mentioned in* Time *magazine:* "Atomic Sub," *Time*, Feb. 26, 1951.

45 *an authorized, fawning biography:* Blair, *The Atomic Submarine and Admiral Rickover.*

47 *"I sometimes get pretty tired of Kansas City"*: Harry S. Truman, "Address in Groton, Conn., at the Keel Laying for the First Atomic Energy Submarine." The speech Truman apologized for was "Address in Springfield at the 32d Reunion of the 3rd Division Association."

48 *"There are many people who have played a role"*: Blair, *The Atomic Submarine and Admiral Rickover*, 5.

48 *Rickover shook the president's hand:* Polmar and Allen, *Rickover,* 190.

49 *"held tenaciously to a single important goal":* Blair, *The Atomic Submarine and Admiral Rickover,* 7.

JANUARY 3, 1961—9:01 PM

51–55 Interviews with Egon Lamprecht.

CHAPTER THREE: IDAHO

57 *A Shoshone legend:* Clark, *Indian Legends from the Northern Rockies,* 193.

58 *"These people are living in the midst of a desert":* Arthur Kleinkopf, Relocation Center Diary, circa 1942, 3.

59 *"no larger than a brick":* "Atomic Power," *New York Times,* Aug. 9, 1945.

62 *WARNING: DO NOT DISTURB*: Stacy, *Proving the Principle,* 73.

63 Electricity flows from atomic energy: ibid., 66.

64 *Rickover broke ground on the* Nautilus *prototype:* Rockwell, *Rickover Effect,* 99. The groundbreaking was apparently conducted without ceremony, and the exact date is lost.

65 *"some damn fool":* Theodore Rockwell, interview with author.

68 *an observant Muslim:* ibid.

68 *fourteen thousand men:* Rickover, "Getting the Job Done Right."

69 *cabinet filled with the ties:* Rayburn, "The Rickover Effect."

69 *"a stupid jerk":* Zumwalt, *On Watch,* 91, 92.

69 *"greasy":* ibid., 95.

69 *"one of those wise goddamn aides":* ibid., 88.

70 *"perhaps more than anyone":* Carter, *Why Not the Best?,* 57.

70 *"a complimentary word to me":* ibid.

70 *Carter absorbed a year's maximum:* ibid., 56.

71 *"The Navy's failure to recognize this":* "Brazen Prejudice," *Time*, August 4, 1952.

72 *Forrestal's psychiatrists found:* Hoopes and Brinkley, *Driven Patriot*, 461.

72 *"Brass hats":* Pearson, "The Washington Merry-Go-Round," July 31, 1953.

74 *Rickover invited Representative Thomas Jefferson Murray:* Polmar and Allen, *Rickover*, 151.

74 *"I haven't experienced real elation many times":* Rockwell, *The Rickover Effect*, 21.

75 *"too cheap to meter":* Lewis Strauss, speech to the National Association of Science Writers, Sept. 16, 1954.

75 *"In a relatively short time":* "Out of the Magic of A-Power: Things to Come," *Newsweek*, Sept. 19, 1960, 69.

76 *"the great organization that developed the atomic bomb":* Musial, *Learn How Dagwood Splits the Atom!*, inside cover.

76 *"My goodness":* ibid., 27.

76 *The radiation is bad:* ibid., 26.

77 *"duck and cover": Duck and Cover*, educational film produced for the United States Civil Defense by Archer Productions, 1951.

78 *"geographical engineering":* Kirsch, *Proving Grounds*, 3.

78 *carve the harbor in the shape of a polar bear:* ibid., 49.

78 *Between 1957 and 1974, the Atomic:* ibid., 6.

79 *"something new under the sun":* "Down to the Sea," *Time*, Feb. 1, 1954.

80 *created by Walt Disney:* Brooke, "Groton Meets Nautilus in a Final Homecoming."

80 *He wore his admiral's uniform:* Polmar and Allen, *Rickover*, 156.

80 *"Be sure and hit it hard, Mrs. Eisenhower":* "Down to the Sea," *Time*, Feb. 1, 1954.

80 *the* Nautilus *rode high in the ocean:* ibid.

80 *As the start-up progressed:* Rockwell, *Rickover Effect,* 186–87.

81 *this had actually happened to Zinn and his EBR-1:* Stacy, *Proving the Principle,* 64.

82 *two men from the office:* Dennis Wilkinson, interview with author.

THE RECOVERY

85–89 Interviews with Egon Lamprecht.

CHAPTER FOUR: THE ARMY

92 *the Army's research and development budget:* Schwartz, *Atomic Audit,* 157.

94 *"an aggregation of photographic memorizers":* Polmar and Allen, *Rickover,* 308.

94 *"No One Loved West Point More":* interview with James Blaine Lampert, son of General James Benjamin Lampert.

94 *his most important tour of duty:* ibid.

94 *"he didn't lose his temper":* Graves, "Engineer Memoirs."

95 *"I understand you want to build nuclear power plants":* Suid, *The Army's Nuclear Power Program,* 10.

96 *"to meet the demands of catastrophe or defense":* Richard Nixon, "Address to the Governors Conference," Lake George, N.Y., July 12, 1954.

96 *"military-industrial complex":* Dwight D. Eisenhower, "Farewell Address," Jan. 17, 1961.

97 *Western Electric congratulated itself: The DEW Line Story,* commemorative booklet prepared by the Western Electric Co., circa 1960.

98 *155,000 kilowatts:* ibid.

98 *"enough to fill the tank cars of a train 65 miles long":* ibid.

98 *"the success of their project":* Morenus, *DEW Line,* 27.

99 *"the far reaching benefits of this program":* Suid, *The Army's Nuclear Power Program,* 8.

100 *"a sub-committee":* Rockwell, *The Rickover Effect,* 193.

101 *at least one of those bushes:* Suid, *The Army's Nuclear Power Program,* 31.

101 *"the snapping of a cable by a rifle bullet":* Alexander, *Atomic Radiation and Life,* 60.

103 *"of utmost significance to the defense": The DEW Line Story.*

103 *"imaginot line":* Letter to the Editor, from Fred E. Breth of Hobart, N.Y., *Time,* Dec. 16, 1957.

104 *"Red Moon over the U.S.": Time,* Oct. 14, 1957.

104 *"if the aggressor's weapon is the ICBM":* "NORAD: Defense of a Continent," *Time,* Nov. 25, 1957.

105 *"quite feasible":* Suid, *The Army's Nuclear Power Program,* 33.

105 *rolling on wheels ten feet in diameter:* ibid., 44.

105 *"a low order atomic explosion":* ibid.

110 *sixty-three separate times:* 19300, 62.

110 *seven separate times:* ibid.

110 *"the central control rod":* McKeown, *Idaho Falls,* 169.

110 *"No. 9 rod has the best over-all operational record":* 19300, 4.

110 *two of the rods were stuck so severely:* ibid., 65.

111 *plugging the hole with an automotive spark plug: ABWR Quarterly Progress Report; SL-1 Operations and Evaluation,* July 15, 1960, 26.

112 *when the boron strips were observed: SL-1 Accident: Atomic Energy Commission Investigation Board Report,* June 1961, 16.

112 *A historic episode:* Rockwell, *The Rickover Effect,* 183–85.

114 *he quietly banned nuclear vessels from visiting large cities:* Polmar and Allen, *Rickover,* 623–25.

114 *"To my surprise, instead of rearing back:* Lilienthal, *The Journals of David E. Lilienthal,* vol. 3, 531.

114 *"The whole reactor game hangs":* Polmar and Allen, *Rickover,* 615.

115 *"power assist":* 19300, 93.

116 *Perform a reactor pump down—procedure No. 54:* ibid., 97–98.

117 *"Pumped reactor water to contaminated water tank":* ibid., 98.

THE BACHELOR PARTY

119–121 Details are from the Miazga memos of Jan. 30, 1961, and July 25, 1962.

CHAPTER 5: THE AIR FORCE

123 *"Our success in weaving the benefits":* Gantz, ed., *Nuclear Flight,* 12.

124 *The power loading of the* Nautilus: James, "The Politics of Extravagance," 158–91.

124 *five hundred times more heat energy:* ibid.

125 *"shitepoke":* Lambright, *Shooting Down the Nuclear Airplane,* 8.

126 *The* Nautilus, *by contrast cost:* Polmar and Allen, *Rickover,* 148.

127 *the engine's plans manacled to his wrist:* Hansen, *Engineer in Charge,* 224.

127 *"any damned publicity":* Blair, *The Atomic Submarine and Admiral Rickover,* 182.

127 *"man of mystery":* "Atom Plane Chief Man of Mystery," *New York Times,* May 22, 1955.

127 *"He is almost unknown":* "Atom Age General: Donald John Keirn," *New York Times,* Dec. 30, 1957.

128 *"the swimming Russian bear":* Rickover, "The Soviet Naval Program."

128 *"does not see a parallel":* "Atom Age General."

128 *"pattern of action that was simply not helpful":* Polenberg, ed., *In the Matter of J. Robert Oppenheimer,* 245. Testimony was given on April 28, 1954.

129 *"I don't challenge his technical judgment":* ibid.

132 *"designing atomic planes of the future":* "First Plane Flies with Operating Reactor Aboard," *New York Times,* Aug. 7, 1956.

132 *In one final strange footnote:* "The Land that Time Forgot," *Flight International,* Aug. 23, 2005.

133 *"no attempt was made to restrict":* Gantz, ed, *Nuclear Flight,* 97.

134 *an estimated total of 4.6 million curies:* Broscious, *Citizens Guide to the United States Department of Energy's Idaho National Laboratory.*

134 *Three Mile Island in 1979:* Broscious, *Citizens Guide,* cites an oft-repeated statistic that Three Mile Island only emitted 15 curies of radioactive iodine as a comparison. While this is true, Three Mile Island also emitted at least 2 million curies of radioactive noble gases according to most estimates. The difference between the planned releases of the HTRE tests and Three Mile Island remains shocking.

136 *through the life of the ANP program: Review of Manned Aircraft Nuclear Propulsion Program,* February 1963, 14.

137 *"why potatoes turn brown when they are fried":* Lambright, *Shooting Down the Nuclear Plane,* 8.

137 *wings on the ocean liner:* James, "The Politics of Extravagance."

138 *"scientific conservatism":* Lambright, *Shooting Down the Nuclear Plane,* 15.

139 *"a nuclear-powered bomber":* "Soviets Flight Testing Nuclear Bomber," *Aviation Week,* Dec. 1, 1958, 27.

139 *"I have only an intuitive feeling":* York, *The Race to Oblivion,* 72.

140 *"There is absolutely no intelligence":* Lambright, *Shooting Down the Nuclear Plane,* 18.

141 *"completely voluntary":* Finney, "Chief of Research on Atomic Plane to Leave Air Force."

141 *175 people:* Gantz, ed., *Nuclear Flight,* 17.

141 *oversaw the efforts of over seven thousand contractors: Review of Manned Aircraft Nuclear Propulsion Program,* 177.

142 *"I have watched":* Brady, "Nuclear Powered Aircraft."

142 *"While there had been substantial progress":* York, *Race to Oblivion,* 68.

143 *"the possibility of achieving":* Message from the President of the United States Relative to Recommendations Relating to Our Defense Budget, March 28, 1961.

CAMP CENTURY

146 *The project was called Iceworm:* All details about the weaponry of Project Iceworm are from Weiss, "Cold War Under the Ice."

147 *"medium-sized Midwestern cities":* Wager, *Camp Century*, 58.

149 *The total cost to the Army for PM-2A:* Suid, *The Army's Nuclear Power Program*, 59.

151 *"One steak or two?":* Wager, "Life Inside a Glacier," 60.

151 *"experience a revival of their spiritual interests":* ibid., 47.

151 *"dream to operate":* Ed Fedol, interview with author.

151 *"there are very few weapons there":* Wager, "Life Inside a Glacier," 4.

CHAPTER 6: THE INVESTIGATION

155 *380 milliseconds:* Entire timeline is from 19311, page IV-27, Figure IV-4, "SL-1 Excursion Summary."

155 *16.5-foot-high tank:* 19300, 10.

156 *26,000-pound vessel:* 19311, II-29.

156 *nine feet and one inch:* 19311, III-36. The exact distance given in the report was "9 feet 1-1/2 inches plus or minus one inch."

156 *85 feet per second:* 19311, IV-27.

156 *four seconds:* ibid.

157 *"Where Idaho Reactor Tragedy Happened":* IFPR, Jan. 5, 1961, morning edition, 1.

157 *"were scheduled to continue a job":* IFPR, Jan. 4, 1961, 2.

157 *"Radiation . . . was at such a high level":* IFPR, Jan. 5, 1961, 1–2.

157 *"there is no radiation danger":* IFPR, Jan. 4, 1961, 1.

158 *"nothing like the explosion of a nuclear bomb":* IFPR, Jan. 5, 1961, 1.

158 *"nuclear runaway . . . a very sluggish reaction":* ibid., 2.

158 *"Is this fear justifiable?":* IFPR, Jan. 9, 1961, home edition, 6.

158 *too high to even be counted by the standard method:* 19302, 57.

159 *Thirteen wild jackrabbits:* ibid.

159 *twenty-eight milk samples:* ibid.

159 *"The reactors and processing plants of the NRTS":* ibid., 55.

160 *one of fourteen American nuclear submarines:* The fourteen submarines at sea at the time of the incident were the *Nautilus, Seawolf, Skate, Swordfish, Sargo, Sea Dragon, Skipjack, Triton, Halibut, Scorpion, Tullibee, George Washington, Patrick Henry,* and the *Robert E. Lee.*

160 *"we are intensely interested":* "Reactor Blast Probed as AEC Officials Arrive," IFPR, Jan. 6, 1961.

161 *"control rod problems": Annual Report to Congress of the Atomic Energy Commission for 1960,* January 1961, 17.

161 *"Atomic Slowdown": Time,* May 19, 1961.

162 *"it could be a serious blow": Annual Report to Congress of the Atomic Energy Commission for 1960,* 19.

163 *"I'm burning up!":* McInroy, "A True Measure of Criticality," 250.

164 *"God gave me permission":* McKee, "Six Families Sue Over LANL Autopsy Harvest," A1.

165 *A. R. Luedecke actually advocated:* Stacy, *Proving the Principle,* 146.

167 *"The head, which was covered by short brown hair": The SL-1 Reactor Accident: Autopsy Procedures and Results,* June 21, 1961, 25.

167 *fifteen to twenty minutes per man:* ibid.

168 *"that fractured his chest and drove a rib":* ibid., 47.

168 *"from the destruction of his viscera":* ibid.

168 *Lushbaugh put Legg:* ibid., 53, Figure 8.

168 *"this reconstruction scene":* ibid., 56.

169 *a blistering 1,500 R/hour on contact:* ibid., 18.

169 *McKinley's left hand:* McKeown, *Idaho Falls,* 130.

169 *"a rapid, sharp dissection":* Autopsy 19, The SL-1 Reactor Accident: Autopsy Procedures and Results, June 21, 1961, 11.

170 *"They were put in stainless steel sinks":* Report prepared by Donald E. Siefert of Oil, Chemical, and Atomic Workers Union Local 2-652.

172 *"From information currently available":* Paul R. Duckworth, memo to file, June 6, 1961.

173 *"We would appreciate your comment":* Allan C. Johnson, memo to C. A. Nelson, Sept. 8, 1961.

173 *"Should there be the slightest inkling":* C. A. Nelson, memo to Allan C. Johnson, Sept. 15, 1961.

173 *"elemental carbon . . . nitro or nitrate groups":* The Stanford Research Institute Report is reproduced in full in 19311, appendix C.

173 *"there was no sabotage":* ibid., C-4.

175 *The ring was blisteringly radioactive:* 19300, 120.

176 *"In the probing of the accident":* "Incident of Anguish," IFPR, Jan. 8, 1961.

176 *"not more than four inches":* the entire procedure as reproduced in the AEC report of June 1961, 120.

178 *blew it up with pentolite:* 19313, 18.

179 *it had apparently rotated 180 degrees:* 19311, III-35.

179 *"Since the rod, in its disconnected position":* 19300, 155.

180 *"almost full withdrawal":* ibid., 6.

180 *"nearly the entire length of the rod":* ibid., 159.

181 *in less than one-tenth of a second:* 19311, III-109.

181 *"Stuck rod, quickly released":* ibid.

182 *"For the last reactor shutdown"*: 19300, 93.

182 *"Although none of these problems"*: 19313, 147.

184 *"was amazed at the damage wrought"*: Miazga memo of July 25, 1962, 18.

184 *"he believed a rod could be slowly withdrawn"*: ibid., 9.

184 *Sergeant Gordon Stolla concurred:* ibid., 13.

184 *Sergeant Robert Honeycutt had actually pulled:* ibid., 8.

185 *"that all examinations"*: ibid., 5.

185 *he would destroy it by withdrawing the central rod:* ibid., 17.

186 *"The assembly of the SL-1 control rod drives"*: 19300, 6.

186 *"The estimated amount of rod withdrawal"*: ibid., 159.

186 *"[It appears] that the operation"*: 19311, IV-1.

187 *"Although we cannot assign the cause"*: *SL-1 Accident: Atomic Energy Commission Investigation Board Report,* June, 1961, v–vi, a reproduction of a May 10, 1961, memo from Curtis A. Nelson to A. R. Luedecke.

THE BURIAL

191 *"I am authorized to assure you"*: Bruce Ruggles, letter to A. R. Luedecke, Jan. 19, 1961.

192 *A myth would grow up in later years:* See McKeown, *Idaho Falls,* 146, for one version of the "floating coffin" incident. Louis Wenzlaff, who was at the graveside, confirms it never happened.

192 *"a burial containing radioactive material"*: Walter L. Parrot, letter to Michigan Department of Environmental Quality, April 10, 1996.

195 *"for potential beneficial occupancy"*: 19311, I-1.

195 *"brushes, dustpans, brooms"*: ibid., II-14.

196 *"surplus facilities"*: E. F. Perry, *Stationary Low Power Reactor No. 1 (SL-1) Accident Site Decontamination & Dismantlement Project,* Oct. 27, 1995, 3.

196 *"clean, compacted soil"*: ibid.

196 *"Through 21 months of field operations"*: ibid., 6.

196 *"a mechanical nibbler"*: ibid. 4.

197 *"leak proof landfill"*: "Leak-proof landfill installed at INEEL," Associated Press, Oct. 28, 2003.

CHAPTER SEVEN: ENTERPRISE

200 *"a persistent and formidable obstacle"*: Zumwalt, *On Watch*, 85.

200 *"I knew that his Division"*: ibid.

201 *more than the entire U.S. civilian industry:* The first eight commercial reactors in the United States were Shippingport, Dresden 1, Yankee, Big Rock Point, Indian Point 1, Humboldt Bay, Elk River, and the CVTR plant in Parr, South Carolina.

201 *Condit remembers frequently using:* Clay Condit, interview with author.

203 *"certain conditions were observed"*: 19313, 146.

203 *"should not permit criticality"*: ibid.

203 *"Such movements should not be"*: ibid.

203 *"The SL-1 had experienced a history"*: ibid., 151.

204 *"The immediate responsibility"*: *Atomic Energy Commission Investigation Board Report*, June 1961 letter of transmittal, paragraph 2, "Responsibility for the incident."

205 *"A mentally unstable operator"*: Ford, *The Cult of the Atom*, 204.

205 *"The Stationary Low-Power Reactor"*: Broad and Wald, "Milestones of the Nuclear Era."

205 *"Investigations following the explosion"*: Mindar, "On a bitterly cold night, the nuclear reactor blew," B6.

206 *"Rumors circulated that the accident"*: Warchol, "50 Years Later, Idaho's Atomic Energy Boom Is a Bust," A1.

207 *813 alumni of SM-1:* Gordon, "SM-1 Nuclear Power Plant to be Deactivated."

208 *"I have very strong memories of those days":* James B. Lampert, letter to Colonel William F. Reilly, April 3, 1975.

209 *"totally convinced the airplane could have been made":* Bob Drexler, interview with author.

209 *"a lot of political by-play":* Donald J. Keirn, interview with Murray Green, Sept. 25, 1970.

209 *"The vacant TAN facilities went up for rent":* Stacy, *Proving the Principle,* 127.

213 *"the thing wrong at Three Mile Island": No Holds Barred: The Final Congressional Testimony of Admiral Hyman Rickover,* 67–68.

213 *"Because I have my representative sitting in the control room":* ibid., 68.

214 *"competence and integrity":* "Three Mile Island Unit 1's Managers Pass Inspection," *Wall Street Journal,* Nov. 23, 1983.

214 *"ruthlessly exploited":* Molotsky, "Rickover's Son Says His Father Was Exploited."

214 *"I don't know how to respond":* ibid.

214 *"shipped off-site to a Department of Energy Facility": Three Mile Island Accident,* U.S. Nuclear Regulatory Commission Fact Sheet, available at www.nrc.gov, 4.

214 *forty-nine casks in twenty-two shipments:* "Waste at INL: Three Mile Island Debris and Dry Storage," Idaho Department of Environmental Quality, http://www.deq.state.id.us/inl_oversight/waste/tmi.cfm.

215 *three carriers, and nine other nuclear-powered surface warships:* Rickover retired on January 31, 1982. The three carriers were the *Enterprise, Nimitz,* and *Dwight D. Eisenhower.* The nine nuclear-powered surface ships were the *Long Beach, Bainbridge, Truxton, California, South Carolina, Virginia, Texas, Mississippi,* and *Arkansas.*

215 *2,300 years of reactor operation without an incident: No Holds Barred,* 55.

216 *"a National Treasure":* ibid., 2.

216 *"breath of fresh air":* ibid., 5.

216 *"We love you, Admiral"*: ibid., 7.

216 *"The Moor has done his duty"*: ibid., 58. Rickover is quoting German playwright Friedrich Schiller's 1783 work *Fiesco,* act 3, scene 4. The line in its original German is "Der Mohr hat seine Arbeit getan, der Mohr kann gehen."

216 *"Ever since the famous Santa Clara County"*: ibid., 15. The case established Fourteenth Amendment protections for corporations.

216 *"The Navy medical staff has certified"*: ibid., 56.

216 *"The only point for wrestling that I know"*: ibid., 75.

216 *"A preoccupation with the so called bottom line"*: ibid., 13.

217 *"After investigating General Dynamics"*: ibid., 20.

218 *"I do not believe that nuclear power is worth it"*: ibid., 70.

220 *"the last word in my personnel jacket"*: Rockwell, *The Rickover Effect,* 386.

220 *"self serving and of limited value"*: Duncan, *Rickover,* xiii.

220 *"your company could be liable to a suit"*: McDowell, "Rickover Losing Fight to Stop a Biography."

220 *"made it clear that he will do everything"*: ibid.

221 *"A technical man himself"*: Rickover, *Eminent Americans,* 246.

221 *that inspectors twice found "dirty"*: Rickover, *How the Battleship Maine Was Destroyed,* 94.

222 *"From his testimony emerges the portrait of an individual who was unfamiliar with his ship"*: ibid., 55.

EPILOGUE

226 *It is more dangerous, visitors learn:* According to the Nuclear Energy Institute's Web site, "In 2005, nuclear's industrial safety accident rate—which tracks the number of accidents that result in lost work time, restricted work or fatalities—was .24 per 200,000 worker hours. U.S. Bureau of Labor statistics show that it is safer to work at a nuclear power plant than in the manufacturing sector and even in the real estate and finance industries."

227 *up to $500 million apiece:* "New Nuclear Plants Get Risk Insurance," *Wall Street Journal*, Aug. 7, 2006.

228 *"quasi-religious":* Brand, "Environmental Heresies."

228 *"We have no time to experiment":* Lovelock, "Nuclear Power is the Only Green Solution," 31.

228 *"the one energy source that is completely domestic":* George W. Bush, "Remarks at Calvert Cliffs Nuclear Power Plant in Lusby, Maryland," June 22, 2005.

229 *the median for all 104 licensed U.S. plants:* all plant statistics from "U.S. Nuclear Reactor List," a spreadsheet provided by the Energy Information Administration, Department of Energy, http://eia.doe.gov/cneaf/nuclear/page/nuc_reactors/operational.xls.

229 *the sudden reduction in demand:* "More Competition for Coal Seen as AEP Returns, 2,200 mw Nuke," *Coal Week*, April 3, 2000.

231 *At the time of Three Mile Island:* Nuclear Energy Institute Document, "U.S. Nuclear Output with Electricity Generation, 1973–2006," www.nei.org.

SOURCES

BOOKS

Acheson, Dean. *Present at the Creation: My Years in the State Department.* New York: W. W. Norton & Co., 1969.

Alexander, Peter. *Atomic Radiation and Life.* London: Penguin Books, 1957.

Barlow, Jeffery G. *Revolt of the Admirals: The Fight for Naval Aviation, 1945–1950.* Washington, D.C.: Naval Historical Center, Department of the Navy, 1994.

Blair, Clay, Jr. *The Atomic Submarine and Admiral Rickover.* New York: Henry Holt & Co., 1954.

Bonny, J. B. *Morrison-Knudsen Company, Inc.: Fifty Years of Construction Progress.* New York: Newcomen Society in North America, 1962.

Broscious, Chuck. *Citizens Guide to the United States Department of Energy's Idaho National Laboratory.* 12th ed. Troy, Idaho: Environmental Defense Institute, November 2005.

Carlisle, Rodney P., with Joan M. Zenzen. *Supplying the Nuclear Arsenal: American Production Reactors, 1942–1992.* Baltimore: Johns Hopkins University Press, 1996.

Carpenter, David M. *NX-2: Aircraft Nuclear Propulsion.* Jet Pioneers of America, 2003.

Carter, Jimmy. *Why Not the Best? Why One Man Is Optimistic About America's Third Century.* Nashville, Tenn.: Broadman Press, 1975.

Clark, Ella E. *Indian Legends from the Northern Rockies.* Norman: University of Oklahoma Press, 1966.

Daugherty, Charles Michael. *City Under the Ice: The Story of Camp Century.* New York: Macmillan, 1963.

The DEW Line Story. Commemorative booklet prepared by the Western Electric Company, circa 1960.

Duncan, Francis. *Rickover: The Struggle for Excellence.* Annapolis, Md.: Naval Institute Press, 2001.

Fawcett, Ruth. *Nuclear Pursuits: The Scientific Biography of Wilfrid Bennett Lewis.* Montreal: McGill-Queen's University Press, 1994.

Ferrell, Robert H. *Off the Record: The Private Papers of Harry S. Truman.* New York: Harper & Row, 1980.

Ford, Daniel. *The Cult of the Atom: The Secret Papers of the Atomic Energy Commission.* New York: Simon & Schuster, 1982.

Friedman, Norman. *U.S. Aircraft Carriers: An Illustrated Design History.* Annapolis, Md.: Naval Institute Press, 1983.

——. *U.S. Submarines Through 1945: An Illustrated Design History.* Annapolis, Md.: Naval Institute Press, 1995.

Fuller, John G. *We Almost Lost Detroit.* New York: Reader's Digest Press, 1975.

Gantz, Kenneth F., ed. *Nuclear Flight: The United States Air Force Programs for Atomic Jets, Missiles, and Rockets.* New York: Duell, Sloan, & Pearce, 1960.

Goodell, Jeff. *Big Coal: The Dirty Secret Behind America's Energy Future.* New York: Houghton Mifflin Co., 2006.

Hamilton, Lee David. *Century: Secret City of the Snows.* New York: G. P Putnam's Sons, 1963.

Hansen, James R. *Engineer in Charge: A History of the Langley Aeronautical Laboratory, 1917–1958.* Washington, D.C.: National Aeronautics and Space Administration, 1987.

Heinl, Robert Debs. *Victory at High Tide: The Inchon-Seoul Campaign.* Nautical & Aviation Publishing Co. of America, 1979.

Hewlett, Richard G., and Francis Duncan. *Atomic Shield, 1947/1952.* Vol. 2 of *A History of the United States Atomic Energy Commission.* University Park: The Pennsylvania State University Press, 1969.

Holl, Jack M. *Argonne National Laboratory, 1946–96.* Urbana: University of Illinois Press, 1997.

Hoopes, Townsend, and Douglas Brinkley. *Driven Patriot: The Life and Times of James Forrestal*. Annapolis, Md.: Naval Institute Press, 1992.

Kirsch, Scott. *Proving Grounds: Project Plowshare and the Unrealized Dream of Nuclear Earthmoving*. New Brunswick, N.J.: Rutgers University Press, 2005.

Kleinkopf, Arthur. *Relocation Center Diary: War Relocation Center, Hunt, Idaho*. Circa 1942.

Lackenbauer, P. Whitney, Matthew J. Farish, and Jennifer Arthur Lackenbauer. *The Distant Early Warning (DEW) Line: A Bibliography and Documentary Resource List*. Prepared for the Arctic Institute of North America, 2005.

Lambright, W. Henry. *Shooting Down the Nuclear Plane: Inter-University Case Program #104*. Indianapolis: Bobbs-Merrill Co., 1967.

Lamprecht, Egon. *The SL-1 Accident: A First Responder's Account*. Self-published, circa 1985.

Lewis, Andrew L. *The Revolt of the Admirals*. Paper submitted to the Air Command and Staff College, Air University, Maxwell Air Force Base, April 1998.

Lewis, Tom. *Divided Highways: Building the Interstate Highways, Transforming American Life*. New York: Viking Penguin, 1997.

Lilienthal, David E. *Atomic Energy: A New Start*. New York: Harper & Row, 1980.

——. *The Journals of David E. Lilienthal*. 3 vols. New York: Harper & Row, 1964.

Manchester, William. *American Caesar: Douglas MacArthur, 1880–1964*. New York: Little, Brown, & Co., 1978.

McFarland, Keith, and David Roll. *Louis Johnson and the Arming of America*. Bloomington: Indiana University Press, 2005.

McKeown, William. *Idaho Falls: The Untold Story of America's First Nuclear Accident*. Toronto: ECW Press, 2003.

Morenus, Richard. *DEW Line: Distant Early Warning, the Miracle of America's First Line of Defense*. New York: Rand McNally & Co., 1957.

Morris, Richard Knowles. *John P. Holland, 1841–1914: Inventor of the Modern Submarine*. Columbia: University of South Carolina Press, 1998.

Musial, Joe. *Learn How Dagwood Splits the Atom!* Comic book published by the Educational Division of King Features Syndicate, 1949. Foreword by General Leslie Groves.

Plastino, Ben J. *Coming of Age: Idaho Falls and the National Engineering Laboratory, 1949–1990.* Chelsea, Mich.: Bookcrafters, 1998.

Polenberg, Richard, ed. *In the Matter of J. Robert Oppenheimer: The Security Clearance Hearing.* Ithaca, N.Y.: Cornell University Press, 2002.

Polmar, Norman, and Thomas B. Allen. *Rickover: Controversy and Genius.* New York: Simon & Schuster, 1982.

The Race for Atomic Power: The Remarkable History of the National Reactor Testing Station Idaho Falls, Idaho. Washington, D.C.: Atomic Heritage Foundation, 2005.

A Review of Criticality Accidents: 2000 Edition. Los Alamos, N.M.: Los Alamos National Laboratory, May 2000.

Rhodes, Richard. *Dark Sun: The Making of the Hydrogen Bomb.* New York: Simon & Schuster, 1995.

——. *The Making of the Atomic Bomb.* New York: Simon & Schuster, 1986.

Rickover, Hyman G. *Education and Freedom.* New York: E. P. Dutton & Co., 1960.

——. *Eminent Americans: Namesakes of the Polaris Submarine Fleet.* Washington, D.C.: U.S. Government Printing Office, 1972.

——. *How the Battleship Maine Was Destroyed.* Washington, D.C.: U.S. Government Printing Office, 1976.

——. *No Holds Barred: The Final Congressional Testimony of Admiral Hyman G. Rickover.* Washington, D.C.: Center for Study of Responsive Law, 1982.

Rickover, Ruth Masters. *Pepper, Rice, and Elephants: A Southeast Asian Journey from Celebes to Siam.* Annapolis, Md.: Naval Institute Press, 1975.

Rockwell, Theodore. *Creating the New World: Stories & Images from the Dawn of the Atomic Age.* Bloomington, Ind.: 1st Books Press, 2004.

——. *The Rickover Effect: How One Man Made a Difference.* Annapolis, Md.: Naval Institute Press, 1992.

Schwartz, Stephen I., ed. *Atomic Audit: The Costs and Consequences of U.S. Nuclear Weapons Since 1940.* Washington: Brookings Institution Press, 1998.

Stacy, Susan. *Proving the Principle: A History of the Idaho National Engineering and Environmental Laboratory 1949–1999.* Washington, D.C.: U.S. Department of Energy, 2000.

Suid, Lawrence H. *The Army's Nuclear Power Program: The Evolution of a Support Agency.* Westport, Conn.: Greenwood Press, 1990.

Verne, Jules. *20,000 Leagues Under the Sea: The Completely Restored and Annotated Edition.* Translated by Walter James Miller and Frederick Paul Walter. Annapolis, Md.: Naval Institute Press, 1993.

Wager, Walter. *Camp Century: City Under the Ice.* Philadelphia: Chilton Books, 1962.

Walker, J. Samuel. *Three Mile Island: A Nuclear Crisis in Historical Perspective.* Berkeley: University of California Press, 2004.

York, Herbert F. *Race to Oblivion: A Participant's View of the Arms Race.* New York: Simon & Schuster, 1970.

Zumwalt, Elmo R., Jr. *On Watch.* New York: Quadrangle, 1976.

ARTICLES AND LETTERS

"American Notes Gratuities." *Time,* June 17, 1985.

"Atom Age General: Donald John Keirn." *New York Times,* Dec. 30, 1957.

"Atom Plane Chief Man of Mystery." *New York Times,* May 22, 1955.

"Atomic Power." *New York Times,* Aug. 9, 1945.

"Atomic Slowdown." *Time,* May 19, 1961.

Barringer, Felicity. "Old Foes Soften to New Reactors." *New York Times,* May 15, 2005.

Brand, Stewart. "Environmental Heresies." *MIT Technology Review,* May 1, 2005.

"Brazen Prejudice." *Time,* August 4, 1952.

Broad, William J., and Matthew L. Wald. "Milestones of the Nuclear Era: A 50-Year Overview." *New York Times,* Dec. 1, 1992.

Brooke, James. "Groton Meets Nautilus in a Final Homecoming." *New York Times,* July 7, 1985.

Capozza, K. L. "The Dew Line: Ditched Drums and All." *Bulletin of the Atomic Scientists,* January/February 2002.

Cowan, George A. "A Natural Fission Reactor." *Scientific American*, July 1976.

Cumings, Bruce. "Spring Thaw for Korea's Cold War?" *Bulletin of the Atomic Scientists*, April 1992.

"Down to the Sea." *Time*, February 1, 1954.

Dufek, Rear Admiral George J. "Nuclear Power for the Polar Regions." *National Geographic*, May 1962.

Eisenberg, Daniel, and Eric Roston. "New Plants on the Horizon?" *Time*, June 20, 2005.

"Extraordinary Atomic Plane: The Fight for an Ultimate Weapon." *Newsweek*, June 4, 1956.

"The Fastest Submarine." *Time*, Sept. 3, 1951.

Finney, John W. "Atom Aides Scan Effect of Blast." *New York Times*, Jan. 5, 1961.

——. "Chief of Research on Atomic Plane to Leave Air Force." *New York Times*, July 31, 1959.

"First Plane Flies with Operating Reactor Aboard." *New York Times*, Aug. 7, 1956.

"For Atomic Defense." *Newsweek*, Jan. 3, 1955.

"Former Critics See the Light." *USA Today*, July 18, 2005.

Gertner, Jon. "Atomic Balm?" *New York Times*, July 16, 2006.

Gordon, Roy. "SM-1 Nuclear Power Plant to be Deactivated, Chief of Army Engineers Announces." Press release, June 2, 1972, Fort Belvoir archive.

Graves, Ernest. "Engineer Memoirs." Interview conducted by Dr. Frank Schubert in 1985, Office of History, U.S. Army Corps of Engineers, Alexandria, Va.

Heppenheimer, T. A. "Heating up the Cold War." *American Heritage*, Fall 1992.

"Historic Achievement Recognition: Shippingport Atomic Power Station, A National Historic Mechanical Engineering Landmark." Program for Presentation Ceremony, May 20, 1980.

Holtz, Robert. "The Soviet Nuclear-Powered Bomber." *Aviation Week*, Dec. 1, 1958.

Horan, John R. "Interview with John B. Horan 8-18-1994." Interview conducted by Burton R. Baldwin and Thomas L. Baccus. Department of Energy Human Radiation Experiment Document #726467.

"If Kingston's Dick Legg Had Died Today . . ." *Saginaw News*, May 16, 1980.

James, Carolyn C. "The Politics of Extravagance." *Naval War College Review*, Spring 2000.

Keirn, Donald J. "Oral History Transcripts: Keirn, Donald J., Major General, United States Air Force (Retired) Interview conducted by Murray Green on September 25, 1970, in Delaplane, Virginia." Provided by the Air Force Historical Research Agency, Maxwell Air Force Base, Ala.

Lampert, James B. Letter to Colonel William F. Reilly, Commander and Director, Facilities Engineering Support Agency, April 3, 1975, Fort Belvoir archive.

"The Land that Time Forgot." *Flight International*, Aug. 23, 2005.

Langston, Jennifer. "Ill-Fated, Nuclear-Powered Airplane Facilities Dismantled." *Idaho Falls Post-Register*, Jan. 27, 1999.

"Leak-proof landfill installed at INEEL." Associated Press, Oct. 28, 2003.

Linn, Edward. "Case History of a Nuclear Tragedy." *Saga*, September 1961.

Lovelock, James. "Go Nuclear, Save the Planet." *Sunday Times*, Feb. 18, 2007.

——. "Nuclear Power is the Only Green Solution." *Independent*, May 24, 2004.

"The Man in Tempo 3." *Time*, Jan. 11, 1954.

McDowell, Edwin. "Rickover Losing Fight to Stop a Biography." *New York Times*, June 26, 1981.

McInroy, James F. "A True Measure of Criticality: The Human Tissue Analysis Program at Los Alamos." *Los Alamos Science*, No. 23, 1995.

McKee, Jennifer. "Six Families Sue Over LANL Autopsy Harvest." *Albuquerque Journal*, March 15, 2001.

McNeil, Donald G., Jr. "Gen. James B. Lampert is Dead; Was a Defense Aide to Johnson." *New York Times*, July 12, 1978.

"Military Rites Held for Blast Victim." *The Kingston (Michigan) Enterprise*, Jan. 27, 1961.

Mindar, Anne. "On a bitterly cold night, the nuclear reactor blew." *Idaho State Journal*, June 11, 2000, reprinted from *Seattle Times*.

Molotsky, Irvin. "Rickover's Son Says His Father Was Exploited." *New York Times*, July 11, 1986.

Moore, Patrick. "Going Nuclear: A Green Makes the Case." *Washington Post*, April 16, 2006.

"More Competition for Coal Seen as AEP Returns 2,200 mw nuke," *Coal Week*, April 3, 2000.

"Nuclear-Equipped Plane Was Tested Over New Mexico in the '50s." *Orange County Star*, March 24, 1987.

"Nuclear Reactor Plane Tested in the '50s, Paper Says." *Toronto Star*, March 24, 1987.

"Out of the Magic of A-Power: Things to Come." *Newsweek*, September 19, 1960.

Parrot, Walter L. Letter to the Michigan Department of Environmental Quality, April 10, 1996. Provided by Richard Peter.

Pearson, Drew. "The Washington Merry-Go-Round," July 31, 1953.

——. "The Washington Merry-Go-Round," Sept. 9, 1953.

——. "The Washington Merry-Go-Round," May 22, 1955.

Rayburn, Kevin. "The Rickover Effect: Speed Grads Remember Working With 'Father of the Nuclear Navy.'" *University of Louisville Magazine*, Winter 2007.

Rickover, Hyman G. "Getting the Job Done Right." *New York Times*, Nov. 25, 1981. Excerpted from a speech Rickover gave to the Columbia University School of Engineering and Applied Science.

——. The Soviet Naval Program." Excerpts from a speech reprinted in *New York Times*, Nov. 13, 1970.

"Roadblock to Progress." *Time*, June 20, 1960.

Ruggles, Bruce. Letter to A. R. Luedecke, General Manager Atomic Energy Commission, Jan. 19, 1961. Provided by Richard Peter.

"Soviets Flight Testing Nuclear Bomber." *Aviation Week*, Dec. 1, 1958.

Sweet, William. "The Nuclear Option." *New York Times*, April 26, 2006.

"Three Mile Island Unit 1's Managers Pass Inspection." *Wall Street Journal*, Nov. 23, 1983.

Voelz, George. "Human Radiation Studies: Remembering the Early Years,

Oral History of Dr. George Voelz, MD." Interview conducted Nov. 28, 1994, Los Alamos, N.M., by Darrell Fischer and Marisa Caputo.

Wager, Walter. "Life Inside a Glacier." *Saturday Evening Post*, Sept. 10, 1960.

Wald, Matthew L. "Ex-Environmental Leaders Tout Nuclear Technology." *New York Times*, April 25, 2006.

Wald, Matthew L., and Heather Timmons. "Approval Is Sought for Reactors." *New York Times*, Sept. 25, 2007.

——. "Much Talk of a Nuclear Renaissance, but so Far Little Action." *New York Times*, March 3, 2006.

Warchol, Glen. "50 Years Later, Idaho's Atomic Energy Boom Is a Bust." *Salt Lake Tribune*, Sept. 10, 2001.

Weiss, Erik D. "Cold War Under the Ice: The Army's Bid for a Long-Range Nuclear Role." *Journal of Cold War Studies*, Fall 2001.

Wolverton, Mark. "Winged Atom." *American History*, February 2003.

SPEECHES

Brady, J. F. "Nuclear Powered Aircraft." Paper presented at the Society of Automotive Engineers National Aeronautic Meeting, Ambassador Hotel, Los Angeles, Calif., Sept. 29–Oct. 4, 1958.

Bush, George W. "Remarks at Calvert Cliffs Nuclear Power Plant in Lusby, Maryland." June 22, 2005.

Eisenhower, Dwight D. "Atoms for Peace." Speech delivered to the United Nations General Assembly, New York, N.Y., Dec. 8, 1953.

——. "Farewell Address." Jan. 17, 1961.

Hafstad, Lawrence. "Atomic Power and Its Implications for Aircraft Propulsion." Speech delivered to the Institute of Aeronautical Sciences, Ambassador Hotel, Los Angeles, Calif., July 22, 1949.

Nixon, Richard. "Address to the Governors Conferece." Lake George, N.Y., July 12, 1954.

Rickover, Hyman G. "Lead Time and Military Strength." Speech delivered to the Society of Business Magazine Editors, Washington, D.C., Jan. 12, 1956.

——. "A National Standard for Education." Speech delivered to the

Twenty-Eighth Annual Meeting of the Southern Governors' Conference, Hollywood-by-the-Sea, Fla., Oct. 3, 1962.

——. "The Soviet Naval Program." Speech excerpted and reprinted in *New York Times*, Nov. 13, 1970.

Strauss, Lewis. Speech to the National Association of Science Writers, Sept. 16, 1954.

Truman, Harry S. "Address in Groton, Conn., at the Keel Laying for the First Atomic Energy Submarine." Speech delivered at the Electric Boat shipyard, Groton, Conn., June 14, 1952.

——. "Address in Springfield at the 32d Reunion of the 3rd Division Association." Speech delivered at the Shrine Mosque, Springfield, Mo., June 7, 1952.

GOVERNMENT PUBLICATIONS (IN CHRONOLOGICAL ORDER)

Survival in the Air Age: A Report by the President's Air Policy Commission. Washington, D.C.: Government Printing Office, January 1948.

Nuclear Power Plant Testing in the IET: Aircraft Nuclear Propulsion Project. General Electric Corporation, May 1953.

Wehman, George. *Preliminary Report of Fission Products Field Release Test-1.* U.S. Atomic Energy Commission, February 1959. (IDO-12006)

Fission Products Field Release Test—1. U.S. Air Force, September 1959. (ANP Doc. No. NARF-59-32T)

Thumbnail Sketch: National Reactor Testing Station. U.S. Atomic Energy Commission, April 1, 1960.

ABWR Quarterly Progress Report: SL-1 Operations and Evaluation. Combustion Engineering, under contract for the U.S. Atomic Energy Commission, July 15, 1960. (ID0-19017)

Annual Report to Congress of the Atomic Energy Commission for 1960. Washington D.C.: U.S. Government Printing Office, January 1961.

John A. Byrnes III, Idaho Nuclear Power Field Office, United States Army. Memo from Leo A. Miazga to E. B. Johnson, Jan. 20, 1961.

SL-1 Reactor Accident on January 3, 1961: Interim Report. Idaho Falls, Idaho:

Combustion Engineering, under contract for the U.S. Atomic Energy Commission, May 15, 1961. (IDO-19300)

SL-1 Accident: Atomic Energy Commission Investigation Board Report. Joint Committee on Atomic Energy, June 1961.

Thumbnail Sketch: National Reactor Testing Station. U.S. Atomic Energy Commission, June 15, 1961.

The SL-1 Reactor Accident: Autopsy Procedures and Results. Los Alamos Scientific Laboratory, June 21, 1961.

SL-1 Recovery Operations: January 3 Thru May 20, 1961. Idaho Falls, Idaho: Combustion Engineering, under contract for the U.S. Atomic Energy Commission, June 30, 1961. (IDO-19301)

Annual Report to Congress of the Atomic Energy Commission for 1961. Washington, D.C.: U.S. Government Printing Office, January 1962.

IDO Report on the Nuclear Incident at the SL-1 Reactor on January 3, 1961, at the National Reactor Testing Station. Idaho Falls, Idaho: U.S. Atomic Energy Commission, January 1962. (IDO-19302)

Islitzer, Norman. *The Role of Meteorology Following the Nuclear Accident in Southeast Idaho.* Idaho Falls, Idaho: U.S. Weather Bureau, May 1962. (IDO-19310)

Final Report of the SL-1 Recovery Operation. Idaho Falls, Idaho: General Electric Co., June 27, 1962. (IDO-19311)

SL-1 Incident (Supplemental Report). Memo from Leo A. Miazga to E. B. Johnson, July 25, 1962.

Additional Analysis of the SL-1 Excursion: Final Report of Progress July through October 1962. Idaho Falls, Idaho: Flight Propulsion Laboratory Department, General Electric Co., Nov. 21, 1962. (IDO-19313)

Review of Manned Aircraft Nuclear Propulsion Program. Comptroller General of the United States, February 1963. (B146759)

Holl, Jack M., Roger M. Anders, and Alice L. Buck. "United States Civilian Nuclear Power Policy, 1954–1984: A Summary History." Washington, D.C.: U.S. Department of Energy, 1986.

Three Mile Island Accident. U.S. Nuclear Regulatory Commission Fact Sheet, available at www.nrc.gov.

Perry, E. F. *Stationary Low Power Reactor No. 1 (SL-1) Accident Site Decontamination & Dismantlement Project.* (Control #1692). Lockheed Martin Technologies, October 27, 1995.

U.S. NRC Fact Sheet: Three Mile Island Accident. U.S. Nuclear Regulatory Commission, Office of Public Affairs, March 2004.

Expected New Nuclear Power Plant Applications. Retrieved from www.nrc.gov.

MEDIA

The China Syndrome. Film directed by James Bridges, 1979.

Duck and Cover. Educational film produced for the United States Civil Defense by Archer Productions, 1951.

Nuclear Pioneers: Creation of the Experimental Breeder Reactor-1. DVD produced by the Atomic Heritage Foundation, 2003.

The Race for Atomic Power: The Remarkable History of the National Reactor Testing Station, Idaho Falls, Idaho. DVD produced by the Atomic Heritage Foundation, 2005.

The SL-1 Accident: Phases 1 and 2. Film produced by the Idaho Operations Office of the U.S. Atomic Energy Commission, circa 1963.

The SL-1 Accident: Phase 3. Film produced by the Idaho Operations Office of the U.S. Atomic Energy Commission, circa 1963.

LEGAL DOCUMENTS

W-31-109 Eng-52. Contract between the United States of America and General Electric, dated May 15, 1946. Obtained from Department of Energy reading room, Richland Operations Office, Richland, Wash.

INTERVIEWS

Skip Bowman

Clay Condit

Earl Daly

Bob Drexler

Ed Fedol

Frank Fogarty

Fitzhugh "Fritz" Fulton

Jane Ann Lalko

James Blaine Lampert

Egon Lamprecht

Harold McFarlane

William McKeown

Oren Painter

Robert Rickover

Theodore Rockwell

Louis Wenzlaff

Eugene "Dennis" Wilkinson

ACKNOWLEDGMENTS

First to friends and family: my wife Susie, my good friends Doug Bennett and Tom Buchanan, and my parents, Ken and Laura Tucker, dedicated readers of early drafts. Thanks also to Liz Stein, Maria Bruk Aupérin, and my agent Frank Scatoni of Venture Literary.

Many thanks to Bill Schalk of Cook Nuclear Power Plant, Skip Bowman and Scott Peterson of the Nuclear Energy Institute, and Susan Connor of the Newport News *Daily Press* for finding for me priceless contemporary accounts of the cancellation of the USS *United States*. Thanks to William McKeown, author of *Idaho Falls,* for sharing with me his copies of various FOIA-obtained documents. Between his redacted copies and mine, I was able to construct almost-whole versions.

Don Miley spent an entire day with me at the Idaho National Laboratory, and referred me to Egon Lamprecht and Frank Fogarty, both neighbors, and both engaging, hospitable witnesses to the dawn of nuclear power. Thanks also to John Walsh and Harold MacFarlane in Idaho, who gave me time out of their busy schedules.

Several men who worked closely with Rickover deserve thanks: Dennis Wilkinson, Theodore Rockwell, and Clay Condit. Bob Drexler, who worked on the nuclear airplane, also has my gratitude, as do Ed Fedol, Oren Painter, and Earl Daly, all proud Army nukes.

Fritz Fulton flew the NB-36H and spoke with me: thank you. James Blaine Lampert and Hester Hill Schnipper, son and daughter of General James Lampert, were generous with their time and their memories of their dedicated father. Robert Rickover was just as kind.

Rick AmRhein, of Valparaiso University's glorious library, helped me every time I asked. Nicole Brooks, Dorothy Riehle, and Janice Parthree all helped me navigate the complexities of my numerous FOIA requests.

Gus Person, resident historian at Fort Belvoir, was enormously helpful, as were Mike Brodhead, U.S. Army Corps of Engineers Office of History, and Cathy Cox, of the Air Force Historical Research Agency.

Finally, a special thanks to those people who spent part of a cold afternoon in Kingston, Michigan, in the library and beside the grave of Richard Legg: Louis Wenzlaff, Jan and Richard Peter, Jim Peter, Shelly Campbell, and Glenna Ford.

INDEX